Microirrigation Systems
Principles and Practices

Microirrigation Systems

Principles and Practices

Ajai Singh PhD
Professor
Department of Water Engineering and Management
Central University of Jharkhand
Ranchi, Jharkhand

CBS

CBS Publishers & Distributors Pvt Ltd

New Delhi • Bengaluru • Chennai • Kochi • Kolkata • Mumbai
Hyderabad • Jharkhand • Nagpur • Patna • Pune • Uttarakhand

Microirrigation Systems
Principles and Practices

ISBN: 978–93–5466–064–1

First Edition: 2022

Published by Satish Kumar Jain and produced by Varun Jain for
CBS Publishers & Distributors Pvt Ltd
4819/XI Prahlad Street, 24 Ansari Road, Daryaganj, New Delhi 110 002, India
Ph: 011–23289259, 23266861, 23266867   Fax: 011–23243014
Website: www.cbspd.com   e-mail: delhi@cbspd.com; cbspubs@airtelmail.in

Corporate Office: 204 FIE, Industrial Area, Patparganj, Delhi 110 092, India
Ph: 011–49344934   Fax: 011–49344935   e-mail: publishing@cbspd.com; publicity@cbspd.com

Branches

- **Bengaluru:** Seema House 2975, 17th Cross, K.R. Road, Banasankari 2nd Stage, Bengaluru 560 070, Karnataka, India
  Ph: +91–80–26771678/79   Fax: +91–80–26771680   e-mail: bangalore@cbspd.com
- **Chennai:** 7, Subbaraya Street, Shenoy Nagar, Chennai 600 030, Tamil Nadu, India
  Ph: +91–44–26680620, 26681266   Fax: +91–44–42032115   e-mail: chennai@cbspd.com
- **Kochi:** 42/1325, 1326, Power House Road, Opposite KSEB, Power House, Ernakulum-682018, Kochi, Kerala, India
  Ph: +91–484–4059061–67   Fax: +91–484–4059065   e-mail: kochi@cbspd.com
- **Kolkata:** 147, Hind Ceramics Compound, 1st Floor, Nilgunj Road, Belghoria, Kolkata 700056, West Bengal, India
  Ph: +91-9096713055/7798394118, 9836841399   e-mail: kolkata@cbspd.com
- **Mumbai:** PWD Shed, Gala No. 25/26, Ramchandra Bhatt Marg, Next JJ Hospital Gate No. 2 Opp. Union Bank of India, Noorbaug, Mumbai-400009, Maharashtra, India
  Ph: +91-22-66661880/89   e-mail: mumbai@cbspd.com

Representatives

• **Hyderabad**	0–9885175004	• **Jharkhand**	0–9811541605	• **Nagpur**	0–9421945513
• **Patna**	0–9334159340	• **Pune**	0–9623451994	• **Uttarakhand**	0–9716462459

Printed at: Rashtriya Printers, Dilshad Garden, Delhi, India.

to

my wife
Punam
and daughter
Anushka

Ashwin B. Pandya

Secretary General

International Commission on Irrigation and Drainage (ICID)
Commission Internationale Des Irrigations et du Drainage (CIID)

Dated: 7th May, 2020

ICID·CIID

FOREWORD

With increasing population and demand of water, the available water resources are getting stretched across the world. As like other parts of the world, agriculture accounts for 70% of the available water and any efficiency improvement at application level has the potential of providing more coverage to newer areas. Also, the microirrigation systems carry many on-farm management advantages like lesser weed growth, prevention of water logging and better economic returns out of the given quantity of water and fertilizer. Microirrigation systems are indeed very suitable for effective demand management options.

The International Commission on Irrigation and Drainage (ICID) is a leading scientific, technical, international, not-for-profit, non-governmental organization. The main mission is to promote sustainable agriculture water management to achieve water secure world free of poverty and hunger through sustainable rural development. ICID is also promoting the technological development and adoption of microirrigation systems and regularly organizes international events on microirrigation, the latest was **9th International Microirrigation Conference** at Aurangabad, India during January 16-18, 2019.

There are continued innovations taking place in these techniques like integration of solar power, drip combined with pivot irrigation systems and control and management technologies coupled with the remotely sensed models. It is necessary that the basic principles of this fast growing technology are understood by the students at under and postgraduate levels. I am sure that this book, which includes the basics of microirrigation systems and recently published research work on development in the domain of micro-irrigation will serve the student community as well as the concerned faculties and practitioners in the field to a great extent by providing up-to-date information in a clear and concise manner with solved numerical examples. The book will also serve for capacity builder for professionals in the field.

I am happy to acknowledge that Professor Ajai Singh has taken pains to add value to the existing knowledge of microirrigation systems. The book is a good blend of field practices and theoretical principles of the discipline. I would like to congratulate Professor Singh for having brought out this textbook with quite useful and valuable information.

I wish all the success to this book.

Ashwin B. Pandya

Central Office: 48 Nyaya Marg Chanakyapuri New Delhi 110021, India
Tel: 91-11-26116837,91-11-26115679;Fax:91-11-26115962
Email: icid@icid.org; Website: www.icid.org

**CONSULTING ENGINEERS
ASSOCIATION OF INDIA**

Creating Value Ethically for Engineers

CEAI Centre, OCF Plot No.2, Pocket 9
Sector B, Vasant Kunj, New Delhi 110070
Tel 91-11-26134644 / 26139658
Email ceai.ceai@gmail.com
Website www.ceai.org.in

CEAI/163/2020   25th April, 2020

FOREWORD

Efficient water use is necessary for sustainable crop production and microirrigation proved to efficiently provide irrigation water and nutrients to the roots of plants, while maintaining high-yield production. Because not all the soil surface is wetted under irrigation, less water is required for irrigation. Microirrigation has become the most valued innovation in agriculture. Higher water application efficiencies are achieved due to reduced soil evaporation, less surface runoff and minimum deep percolation. Microirrigation methods are also being seen as a potential means to enhance water use efficiency in agricultural field.

The knowledge about microirrigation, especially drip system, and its design principles is essential for the students of agricultural/civil engineering and professionals of agriculture and water resources engineering. I am sure that this book *Microirrigation Systems: Principles and Practices*, which includes the basics of microirrigation systems recent published research work on designing of drip irrigation system, will serve the student community as well as faculties to a great extent by providing up-to-date information in a clear and concise manner.

I would like to compliment Professor Ajai Singh for having brought out this textbook with quite useful and valuable information.

Ajay Pradhan, PhD
Vice President – CEAI
&
President & CEO
C2S2 Pvt Ltd.
Email: ajay.pradhan@c2s2.in

PREFACE

Adoption of microirrigation system can be a panacea in irrigation-related problems and can boost the saving of on-farm irrigation water. In this technology, the cropped field is irrigated in the close vicinity of root zone of crop. It reduces the water loss occurring through evaporation, conveyance and distribution. Therefore, high water use efficiency can be achieved. The unirrigated rainfed cropped area can be increased with this technology and potential source of food production for the benefit of country's food security could be augmented. The Government of India has been considering rapid promotion of use of plastics in agriculture and microirrigation as a major step in improving overall horticultural crop yields and water use efficiency. The microirrigation has gained considerable growth in the country due to financial assistance provided by the centrally sponsored subsidy scheme. This book is designed as a professional textbook with basic and updated information and its aim is to meet the needs of the students of agricultural/civil engineering undergraduate and postgraduate degree program. The practicing engineers, agricultural scientists working in the field of agricultural water management and officers of Agriculture and Horticulture Departments of State Government will find this book very useful. In this book, the basics of microirrigation, types, components, design, installation, operation and maintenance are presented. Efforts have been made to incorporate the recent published works in peer reviewed journals at appropriate place. Numerical problems and examples have been added to emphasize the design principles and make the understanding of the subject matter.

The author is grateful to many individuals and organizations for the assistance provided at different stages of preparation of manuscript and supply of useful literature. I am grateful to Professor RP Singh from GBPUAT, Pantnagar and Professor KN Tiwari, AgFE Department, IIT, Kharagpur for their love, affection and guidance. I thank Ms Pratibha Kumari, Ms Nity Tirky, Research Scholars, and Mr Rajnish Kumar, Ms Fakeha Parween, M. Tech students of Department of Water Engineering and Management for their support in framing numerical and making artwork.

The author is indebted to many people and institution/organizations for providing the material and issuing the permission under Copyright Act for inclusion in this textbook. I wish to acknowledge the support received from the Food and Agricultural Organization, Arizona Cooperative Extension, Arizona

State University and Dr Richard John Stirzaker of CSIRO, Division of Land and Water, Australia for providing the figures and technical details during my first book and same has been used in this book also. The author has tried to acknowledge the every source of information, any omission is inadvertent. This can be pointed out and will be included in the book at appropriate time.

I feel profound privilege in expressing my heartfelt reverence to my parents, brothers and sister, in-laws for their blessings and moral support to achieve this goal. Last but not the least I acknowledge with heartfelt indebtness, the patience and the generous support rendered by my wife, Punam and daughter Anushka who always allowed me to work continuously with smile on their face. Author would welcome suggestions from readers to improve the text of the book at ajai.singh@cuj.ac.in.

Ajai Singh

CONTENTS

1

INTRODUCTION

National Mission on Microirrigation (NMMI) had a clear vision to promote microirrigation as a thrust area and hence the area under this technology was increased from 3.09 Mha in 1992 to 6.14 Mha in 2012. Under NMMI, some of the States like Bihar, Karnataka, Odisha, Rajasthan and Sikkim achieved more than 90% of the set targets (physical and financial) whereas Andhra Pradesh, Chattisgarh, Gujarat, Haryana, Maharashtra and Tamil Nadu achieved more than 70% of the target. Out of a total 140.13 Mha of sown area, India's net irrigated area is 68.38 Mha, while 71.74 Mha are unirrigated. To bridge this gap, the government launched the Pradhan Mantri Krishi Sinchayee Yojna (PMKSY) in 2015–2016 by combining ongoing schemes. The PMKSY has two distinctive slogans for water management: (i) *Har Khet ko Pani*—extension of irrigation cover and (ii) per drop more crop—improving water use efficiency which aimed at boosting investment in irrigation and improving efficiency of water use. Andhra Pradesh has covered as 186,444 ha area under microirrigation followed by Karnataka where total area coverage under microirrigation is reported around 165,000 ha.

Indian agriculture continues to be the backbone of our society and it provides livelihood to nearly 50% of our population. Continuous innovation and efforts towards productivity, pre and post-harvest management, processing and value-addition, use of technology and infrastructure creation is essential for Indian agriculture. India's share in global exports of agriculture products has increased from 1% a few years ago, to 2.2% in 2016. India occupies a leading position in global trade of agricultural products. India has remained consistently a net exporter of agri products, touching ₹2.7 lakh crore exports and imports at ₹1.37 lakh crore in 2018–19 as report of Economic Survey for 2018–19. The share of India's high value and value added agri produce in its agri export basket is less than 15% as compared to 25% in US and 49% in China. Much scope is still there to improve our agri export in next 5–10 years.

Land and water are two most important resources for any activity in the field of agriculture. Water is such a natural resource which cannot be replenished and its demand is increasing alarmingly. The country is endowed with many perennial and seasonal rivers. The river system which constitutes 71% of water resources, is concentrated in 36% of geographical area. Most of the agricultural fields are irrigated by use of underground water for assured irrigation. Rainfall is a source for water for rainfed agriculture. In present times, the water resources play a very significant role in development and constitute a critical input for economic planning in developing countries.

Large investments and phenomenal growth in the irrigation sector, the returns from irrigated systems in terms of crop yield, farm income and cost recovery are disappointing. Apart from that, there are additional problems of increase in soil salinity, water logging and social inequity. There is a large gap between the development and utilization of irrigation potential created. If we go back to the past of development of drip irrigation system, we find that Davis (1974) used subsurface clay pipes with irrigation and drainage systems in an experiment. Irrigation of plants through narrow openings in pipes can also be traced back to greenhouse operations in the United Kingdom in the late 1940s (Davis, 1974). Blass (1964) observed that a tree near a leaking faucet exhibited a more vigorous growth than other trees in the area. He worked on it and developed the current form of drip irrigation technology and got patented.

The availability of low cost plastic pipe for water delivery lines was one of the reasons which helped to spread the application of drip irrigation systems. Gradually, the area under drip irrigation increased throughout the world especially in countries where water was a scarce resource. Although drip irrigation systems are considered the leading water saving technologies in irrigated agriculture, their adoption is still low. Most of the drip irrigated area is concentrated in Europe and the America.

Asia has the highest area under irrigation (193 million ha, which is 69% of the total irrigated area), but has very low area of 1.8 million ha (<1%) under drip irrigation. In some countries, such as Israel and Jordan, where water availability limits crop production, drip irrigation systems irrigate about 75% of the total irrigated area. In India, it accounts for 2.3% of the total irrigated area (62.3 million ha). While the ultimate potential for drip irrigation in India is estimated at 27 million ha.

Drip irrigation, like other irrigation methods, will not fit every agricultural crop, specific site or objective. Presently, drip irrigation has the greatest potential, where (*i*) water and labor are expensive or scarce, (*ii*) water is of marginal quality, *viz.* saline, (*iii*) soils are sandy, rocky or difficult to level, (*iv*) steep slopes and undulated topography, and (*v*) high value crops are produced. The principal crops under drip irrigation are commercial field crops (sugarcane, cotton, tobacco, etc.) horticultural crops—fruit and orchard crops, vegetables, flowers, spices and condiments, bulb and tuber crops, plantation crops and silviculture/forestry plantations. This method of irrigation continues to be important in the protected agriculture, *viz*. greenhouses, shade nets, shallow and walking tunnels, etc. for production of vegetables and flowers.

Drip irrigation is also used for landscapes, parks, highways, commercial developments and residences. Undoubtedly, the area under drip irrigation will continue to increase rapidly as the amount of water available to agriculture declines and the demands for urban and industrial use increase. Drip irrigation is also one of the techniques that enable growers to overcome salinity problems that currently affect 8.0 million ha area in India. As this area increases, so too will the use of drip irrigation to maintain crop production. In addition, because growers are looking to reduce cost of production but at the same time improve crop quality, the improved efficiency provided from drip irrigation technology will become increasingly important.

Apart from drip irrigation systems or we can say microirrigation systems; there are several indigenous low pressure low volume irrigation systems. For example, there are pitcher methods, low cost drip irrigation systems and bamboo-based drip irrigation systems. Under pitcher methods, earthen pots with a hole on the bottom are placed in a ring basin made around the plants. Pots can be of 10–20 L capacity and need to be refilled when gets empty. Pitcher method is used for irrigation of small area and where energy availability is irregular.

A simple low-cost drip irrigation system uses plastic pipes laid on the surface to irrigate vegetables, field crops and orchards. Water is delivered through the small holes made in the hose. It can consist of a 20 L bucket with 30 m of hose or drip tape connected to the bottom of the tank. The bucket is placed at 1–2 m above the ground so that gravitation head can create sufficient water pressure to ensure watering of the crops. Maintenance usually involves repairing the leakages in the pipes and joints and clearing blockages. Bamboo drip irrigation method is used to divert the part of the flow of hillside stream by using a hallow bamboo in place of earthen channel. Government of India is also providing subsidy to farmers in order to boost-up the adoption of this drip irrigation technology among the farmers. Drip irrigation techniques has always been considered to be adopted only in water scarce area but, I strongly believe that time has come to adopt and propagate this system in areas endowed with abundant supply of water just to make sure the sustainable development of agriculture.

■■

MICROIRRIGATION SYSTEMS

Microirrigation is defined as slow application of water above or below the soil surface near the vicinity of plant roots. Water is applied in the form of drops, sprayed over the land surface or in small continuous stream through fixed applicator near the plants. The basic concept underlying the microirrigation method is to supply the amount of water needed by the plant within a limited volume of the soil rather than wetting the whole area. Microirrigation refers to low-pressure irrigation systems that spray, mist, sprinkle or drip. Water is applied close to the plants so that only part of the soil in which the roots grown is wetted, unlike surface and sprinkler irrigation, which involves wetting the whole soil profile. In this method of irrigation, a dense root system is developed in the zone adjacent to the dripper, resulting in direct and therefore more efficient water use by the plant. A network of pipes and a large number of drippers are required in the field because the discharge of a dripper is small (2 to 8 lph). With microirrigation, water applications are more frequent than with other methods and this provides a very favorable moisture level in the soil in which plants can flourish. This system is best suited to the area having scanty rainfall or poor quality irrigation water is being used. The low volume irrigation systems are also suitable for almost all orchard crops, plantation crops and most of the vegetable crops. Microirrigation has gained attention during recent years because of its potential to increase yields and decrease water, fertilizer, and labor requirements if managed properly.

2.1 Types of Microirrigation System

The microirrigation system is classified based on the installation method, emitter flow rate, wetted soil surface area or the mode of operation (Fig. 2.1). Types of the microirrigation systems are briefly described below.

2.1.1 Drip Irrigation

Drip irrigation applies water directly to the soil surface and allows the water to dissipate under low pressure in the form of drops. A wetted profile develops in the plant's root zone beneath each dripper. The shape depends on soil characteristics, but often it is onion-shaped. Ideally, the area between rows or individual plants remains dry and receives moisture only from incidental rainfall. In this system, the emitters and laterals are laid on the land surface. It has been primarily used on widely spaced plants, but can also be used for row crops. Generally, discharge rates are less than 12 lph for single outlet point-source emitter and less than 12 lph per meter of lateral for line-source emitters. Advantages of this system include the ease of installation, changing and cleaning the emitters and measuring individual emitter discharge. Often the terms drip irrigation and trickle irrigation are considered synonymous. It is suitable for establishing the forestry plantations under wasteland development program. Still we are applying drip irrigation to water scarce area to grow the crops.

(A)   (B)
Fig. 2.1: Emitter connected with the lateral (Jain irrigation systems)

2.1.2   Subsurface System

It is a system in which water is applied slowly below the surface through the line-source emitters. The water is applied through emitters with discharge rate generally in the range of 0.6 to 4 lph. A thin-walled drip line has internal emitters glued together at pre-defined spacings within a thin plastic distribution line. The drip line is available in a wide range of diameters, wall thickness, emitter spacing, and flow rates. The emitter spacing is selected to closely fit plant spacing for most row crops. Drip lines are buried below the ground and therefore called subsurface drip irrigation systems (Fig. 2.2). Burial of the drip line is preferable to avoid degradation from heat, ultraviolet rays and displacement from strong winds. These systems are used on small fruits and vegetable crops. Advantages of subsurface system include freedom from anchoring of the lateral lines at the beginning and removing them at the end of the growing season, little interference with cultivation and possibly a longer operational life.

Fig. 2.2: Subsurface drip irrigation systems (*source:* Peter Thorburn, www.askgillevy.com)

2.1.3 Bubbler

Bubblers are very similar to the point source online emitters in shape but differ in performance. In this system, the water is applied to the soil surface in a small stream or fountain from an opening with a point-source. The discharge rate is usually greater than surface or subsurface drip irrigation but usually less than 225 lph. A small basin is required to control the distribution of water. Advantages of bubbler system are reduced filtration, maintenance or repair and energy requirements as compared with other microirrigation systems. Larger size lateral is used with this system to reduce the pressure loss associated with a high discharge rate. The bubbler heads are used in planter boxes, tree wells, or specialized landscape applications where deep localized watering is preferable (Fig. 2.3). High irrigation application efficiency up to 75% can be achieved with total control of the irrigation water. Another advantage is that the entire piping network is buried, so there are no problems in field operations. Associated disadvantages are not supplying the small water flows as in other microirrigation systems. It is not possible to achieve a uniform water distribution over the tree basins in sandy soils with high infiltration rates.

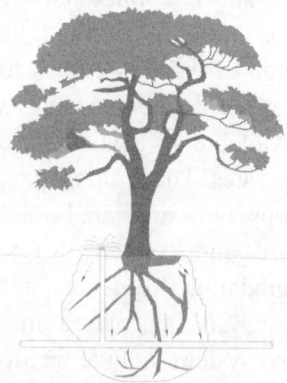

Fig. 2.3: Bubbler placed near the tree and fitted with underground pipelines

2.1.4 Microsprinkler

This is a combination of sprinkler and drip irrigation. Water is sprinkled around the root zone of plants with a small sprinkler working under low pressure. Water is given only to the root zone area as is in the case of drip irrigation but not to the entire ground surface as done in the case of sprinkler irrigation method. Depending on the water throw patterns, the microsprinklers are referred to as minisprays, microsprays, jets, or spinners (Fig. 2.4). The sprinkler heads are external emitters individually connected to the lateral pipe typically using spaghetti tubing, which is very small (1/8 inch to 1/4 inch) diameter tubing. The sprinkler heads can be mounted on a support stake or connected to the supply pipe. Microsprinklers are desirable because fewer sprinkler heads are necessary to cover larger areas. Microsprinklers require 35 kPa to 300 kPa of pressure for operation. Discharge rates usually vary from 15 lph to 200 lph. Microsprinkler system is less likely to clog than a drip irrigation system, but water losses due to wind drift and evaporation are greater.

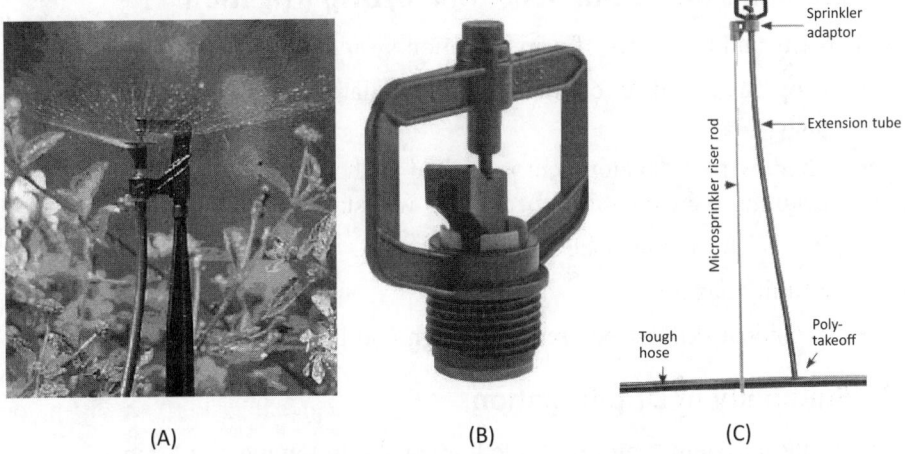

(A) (B) (C)

Fig. 2.4: Microsprinklers (Jain irrigation systems)

2.1.5 Pulse System

Pulse system uses high flow rate emitters and consequently has a shorter water application time. Pulse systems have application cycles of 5, 10, or 15 minutes in an hour, and flow rates for pulse emitters are 4 to 10 times larger than the conventional surface drip irrigation system. The primary advantage of this system is a possible reduction in the clogging problems.

2.2 Advantages of Microirrigation

The main advantages of micro/drip irrigation systems are:

- Water saving

- Enhanced plant growth and yield
- Uniform and better quality of produce
- Efficient and economic use of fertilizers
- Less weed growth
- Possibility of using saline water
- Energy saving
- Can be automated
- Improved production on undulating land conditions
- No soil erosion
- Flexibility in operation
- Labor saving
- No land preparation
- Minimum disease and pest infestation.

2.3   Problems and Demerits of Micro/Drip Irrigation

The demerits and problems of drip irrigation system are given below:

- Clogging of drip emitters by particulate, chemicals and biological materials
- Shallow root development is limited to the wetted portion of root zone, resulting reduced ability of trees to withstand against high winds
- Persistent maintenance requirement
- Salinity hazards
- Technical skill is required for design and installation.

2.4   Suitability of Drip Irrigation

Drip irrigation systems can be adopted under the following conditions:

2.4.1   Crops

Drip irrigation is most suitable for vegetables, fruits, sugarcane, and cereal crops except paddy. The high value crops such as fruit crops give early recovery of capital investment on installation of a microirrigation system. These systems are also suitable for plantation crops such as coconut, coffee, cardamom, cumin, citrus, grapes and mango. Close growing crops will require more investment, otherwise for widely spaced crops, these systems can be easily installed.

2.4.2   Slopes

Drip irrigation is adaptable to any cultivable slope. Normally, the crops and laterals are planted along contour lines. This practice minimizes the change in emitter discharge due to change in land elevations.

2.4.3 Soil

Drip irrigation is suitable for most of the soils. For example, on clay soils, irrigation water should be applied slowly to avoid ponding and runoff. On sandy soils, higher emitter discharge will be appropriate to ensure lateral movement of the water into the soil. It can be applied to irrigate crops grown on undulating land topography and slopes where the depth of soil is limited.

2.4.4 Irrigation water

One of the major problems with drip irrigation is emitters clogging. All emitters have very small openings ranging from 0.2–2.0 mm in diameter and these can be clogged with the use of dirty water. Thus, it is essential to install filters for irrigation water to be free from sediments. Microsprinklers can eliminate problem of clogging to a certain extent. Clogging may also occur if the water contains algae, fertilizer deposits and dissolved chemicals which precipitate such as calcium and iron.

Filtration may remove some of the materials. Drip irrigation is also suitable for poor quality water (saline water). Supplying water to individual plants also means that the method can be efficient to increase the water use efficiency and thus most suitable where water is a scarce resource.

2.5 Quantitative Approach of Wetting Patterns

Due to the manner in which water is applied by a drip irrigation system, only a portion of the soil surface and root zone of the total field is wetted unlike surface and sprinkler irrigation systems. Water flowing from the emitter is distributed in the soil by gravity and capillary forces creating the contour lines similar to onion shape. The exact shape of the wetted volume and moisture distribution will depend on the soil texture, initial soil moisture, and to some degree, on the rate of water application. Figures 2.5 and 2.6 show the effects of changes in discharge on two different soil types, namely sand and clay. The water savings that can be made using drip irrigation are the reductions in deep percolation, surface runoff and evaporation from the soil. It is evident from the Fig. 2.7 that soil moisture content in the soil always remains at or around the field capacity in drip irrigation, whereas in sprinkler and surface irrigation methods, crops face overirrigation and water stress during certain period.

In the line source type of drip irrigation system where the emitters are spaced very closely, individual onion patterns creates a continuous moisture zone. The knowledge about the wetting patterns under emitters is essential in selecting the appropriate spacing of the emitters. Distance between emitters and emitter flow rates must match to the wetting characteristics of the soil and the amount and timing of water to be supplied to meet the crop needs.

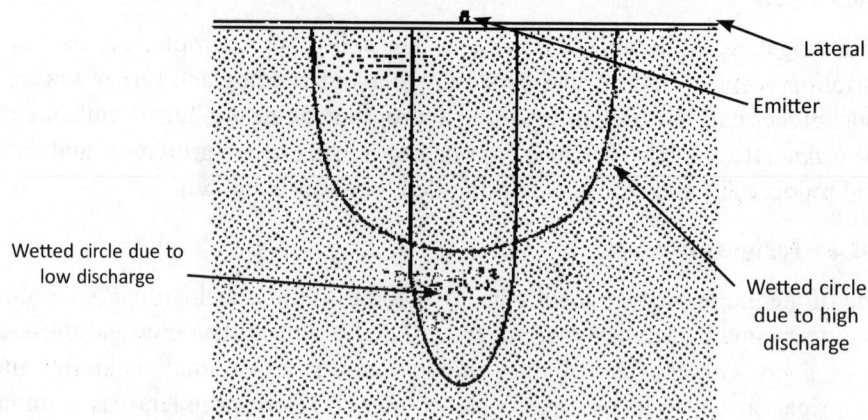

Fig. 2.5: Wetting patterns for sandy soils with high and low discharge rates emitters

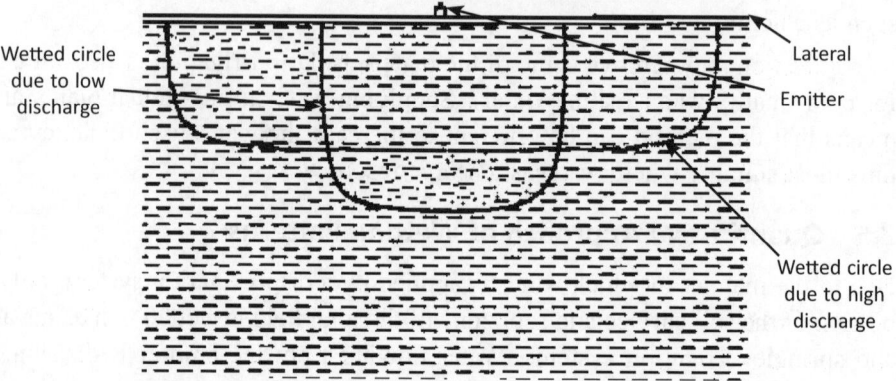

Fig. 2.6: Wetting patterns for clay soils with high and low discharge rates emitters

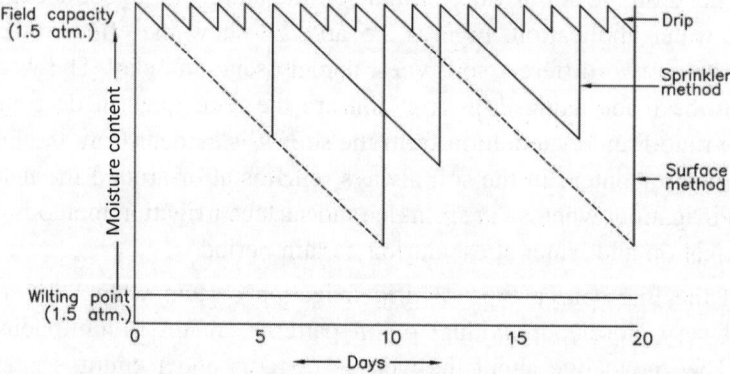

Fig. 2.7: Moisture availability for crops in different irrigation methods

Wetting patterns around a surface drip line and a subsurface drip line just after an irrigation show that how the soil moisture content varies around the drip line (Fig. 2.8 A and B). Agriculture and Natural Resources division of University of California provided the colored photographs which clearly depicts the advancement of soil moisture with time. The wettest soil is near the dripper and the driest is at the periphery of the wetted pattern. Root distribution around drip lines also reflects wetting patterns. The roots of a row crop are highly concentrated near the zone of wettest soil if the drip line placement coincides with the plant row (Hanson and May, 2007).

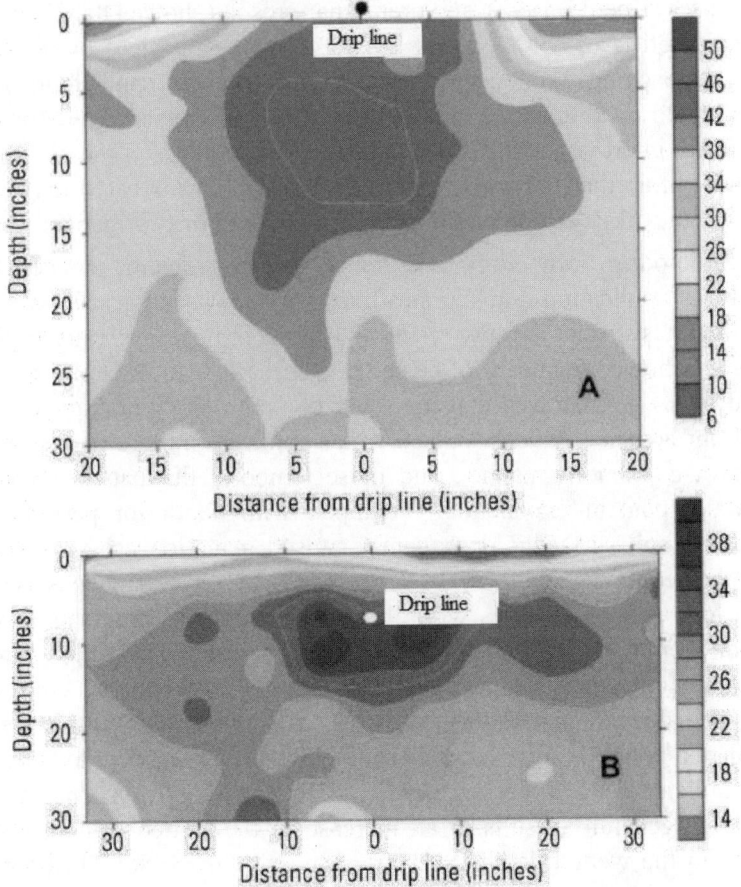

Fig. 2.8: Soil moisture patterns around drip lines for (A) a surface drip line and (B) a subsurface drip line. Units in the color scale indicate % volumetric moisture content

Under drip irrigation, the ponding zone that develops around the emitter is strongly related to both the application rate and the soil properties. The water application rate is one of the factors which determine the soil moisture regime around the emitter and the related root distribution and plant water uptake patterns (Coelho and Or, 1996, 1999). Drip irrigation systems generally consist of emitters that have discharge varying from 2.0 to 8.0 lph. In semi-arid climates, crop water use during the summer can be 6 to 8 mm/d with water supplied two or three times a week. When the water application is exactly equal to the plant water need, then also, part of the water may not be used by the plant and it would most likely leach below the root zone. Therefore, lowering the emitter discharge to as close as possible to the plant water uptake rate can improve irrigation efficiency.

Recently, microdrip irrigation systems have been developed that provide emitter discharges of 0.5 lph. These systems have been studied most intensively in greenhouses (Koenig, 1997), and preliminary results showed that they reduced water consumption of tomato plant by 38%, increased yield by 14 to 26%, and reduced leaching fraction by 10 to 40%. In a recent application on sweet corn under field conditions, Assouline et al. (2002) have shown that microdrip irrigation may improve yield, reduce drainage flux, and affect the water content distribution within the root zone, especially through an increased drying of the 0.60 to 0.90 m soil layer compared with conventional drip irrigation.

The microdrip technology still raises some problems concerning the uniformity of application and the steadiness of the discharges. However, soil moisture regimes similar to those resulting from continual low water application rates can be achieved by means of pulsed drip irrigation. Infiltration experiments on a sandy loam soil showed that the water content distribution and the rate of wetting front advance under a pulsed water application were similar to water applied in a continuous manner, and those temporal fluctuations in flux and in soil water content exponentially damped with depth for periodic pulses applied at the soil surface. Consequently, pulsed irrigation using conventional drip emitters could be one way of creating the water regime observed with continual low application rates while bypassing technical problems associated to microdrip emitters. The relationships between water application rates, soil properties, and the resulting water distribution for conventional emitters (2.0 lph) are well documented. The wetting patterns during application generally consist of two zones—(i) a saturated zone close to the emitter, and (ii) a zone where the water content decreases toward the wetting front. Increasing the emission rate generally results in an increase in the wetted soil diameter and a decrease in the wetted depth (Schwartzman and Zur, 1986; Ah Koon et al., 1990). In microdrip irrigation, field observations seem to indicate that there is no saturated zone and that the wetted soil volume is greater compared with that for conventional emitter discharges (Koenig, 1997). The relationship between the water application rate and the resulting water content distribution is complex because it is a three-dimensional outcome related to soil properties and crop uptake characteristics. Therefore, a quantitative representation of the flow processes by means of a simulation model could be beneficial in studying the effects of emitter discharge on the water regime of drip irrigated crops.

Many attempts have been made to determine water movement and wetting pattern under drip emitters using mathematical and numerical models. The

Richards equation, formulated by Lorenzo A. Richards in 1931, describes the movement of water in unsaturated soils. It is a non-linear partial differential equation, which is often difficult to approximate. Partial differential equations are a type of differential equation which formulates a relation involving unknown functions of several independent variables and their partial derivatives with respect to those variables. Ordinary differential equations usually model dynamical systems whereas partial differential equations are used to model multi-dimensional systems.

In Darcy's law, if there is no gradient, i.e. H = 0, there is no flow. Darcy's law was developed for saturated flow in porous media; to this Richards applied a continuity requirement and obtained a general partial differential equation describing water movement in unsaturated soils. The Richards' equation is based solely on Darcy's law and the continuity equation. Therefore it is strongly physically based, generally applicable, and can be used for fundamental research and scenario analysis. Richards equation can be written in a number of forms such as the water content, mixed water content, capillary head form and the head form. Darcy–Buckingham law extended Darcy's law for unsaturated flow condition and proposed the following equation–

$$q = -K\left(\psi_m\right)\left[\frac{\partial \psi_m}{\partial z} + \frac{\partial \psi_z}{\partial z}\right] \qquad \qquad ...(2.1)$$

Ψ_m is the *matric potential energy* or the *matric potential* is the portion of the water potential that can be attributed to the attraction of the soil matrix for water. The matric potential used to be called the *capillary potential*, because, over a large part of its range, the matric potential is due to capillary action akin to the rise of water in small, cylindrical capillary tubes.

In the above Eqn. 2.1, $\frac{\partial \psi_z}{\partial z}$ is gradient of gravimetric potential and $\frac{\partial \psi_m}{\partial z}$ is gradient of matric potential wrt z. $\frac{\partial \psi}{\partial z}$ is considered as 1. Here, $K(\Psi_m)$ is hydraulic conductivity of matric potential. The equation becomes

$$q = -K\left(\psi_m\right)\left[\frac{\partial \psi_m}{\partial z} + 1\right] \qquad \qquad ...(2.2)$$

The equation can be used for steady state condition, i.e. flow entering into a vertical column of soil will be equal to flow leaving the column. For unsteady state condition in natural condition, we used to have transient flow. We need to write the continuity equation in the given form.

$$\frac{\partial \theta}{\partial t} = -\frac{\partial}{\partial z}[K\left(\psi_m\right)\left[\frac{\partial \psi_m}{\partial z} + 1\right] \qquad \qquad ...(2.3)$$

That means volumetric water content wrt to time (LHS) should be equal to negative of flow rate wrt space. Therefore Richards equation can be written in the form of capillary head.

Richards equation can be written in a number of forms such as the water content, mixed water content, capillary head form and the head form. In one-dimension, the mixed water content form, because it mixes the water content θ with the capillary head $\psi(\theta)$, is:

$$\frac{\partial \theta}{\partial t} = \frac{\partial}{\partial z}\left[K(\theta)\left(\frac{\partial \psi(\theta)}{\partial z} - 1\right)\right] \qquad ...(2.4)$$

The Eqn. 2.4 can be written as a function of water content by introducing Chain Rule.

$$\text{If } D(\theta) = K(\theta)\frac{\partial \psi(\theta)}{\partial \theta}$$

Multiply with $\partial\theta/\partial z$ on both the sides in above equation

$$D(\theta)\frac{\partial \theta}{\partial z} = K(\theta)\frac{\partial \psi(\theta)}{\partial \theta}\frac{\partial \theta}{\partial z}$$

From Eqn. 2.4, we can write

$$\frac{\partial \theta}{\partial t} = \frac{\partial}{\partial z}\left[D(\theta)\frac{\partial \theta}{\partial z} - K(\theta)\right] \qquad ...(2.5)$$

Here, $D(\theta)\frac{\partial \theta}{\partial z}$ is termed soil water diffusivity.

In Eqn. 2.4, the first term in parentheses captures the effects of capillarity, while second term in parentheses represents the effect of gravity-driven flux. In uniform soils the water content or mixed water content forms of Richards equation are valuable because water content is a continuous variable. However, in nature soils are seldom uniform over significant length scales and layered soils are ubiquitous. In layered soils, the water content is discontinuous across layer interfaces because of unique unsaturated capillary head relations in the different soil layers (Assouline, 2013).

Rather, the capillary head (ψ) is continuous, and it is better to write the Richards equation with capillary head as the dependent variable and evaluate the moisture content in terms of ψ, $\theta = (\psi)$. The capillary head function is $\psi(\theta)$, and the specific moisture capacity is defined as $\partial/\partial\psi$. Analytical solutions of Richards equation exist only for simplified cases, so most practical situations require a numerical solution in one- two- or three-dimensions, depending on the problem and complexity of the flow situation. The one-dimensional solution methodology developed by Celia et al. (1990), which uses modified Picard iterations to improve mass conservation, has become the standard numerical approach. It remains essentially the method that is used in many production codes including the USDA Hydrus-1D Richards equation solver (Simunek et al., 2005). The numerical solution of the Richards equation remains computationally expensive and in certain circumstances, unreliable. Farthing and Ogden (2017) have very well documented the existing improved solution methodologies for solving Richards equation.

Under drip irrigation, we have already discussed that only a portion of the horizontal and cross-sectional area of the soil is wetted. The percentage wetted area as compared with the entire field covered with crops, depends on the volume and rate of discharge at each emitter, spacing of emitter and the type of soil being irrigated. For widely spaced crops, the percentage wetted area should be less than 67% in order to keep the area between the rows relatively dry for cultural practices. Low value of percentage wetted area also reduces the loss of water due to evaporation and involves less cost. For closely spaced crops such as vegetables with rows and laterals spaced less than 1.8 m, percentage wetted area often approaches 100% (Keller and Bliesner, 1990). Several efforts have been made to estimate the dimensions of the wetted volume of soil under an emitter.

Schwartzman and Zur (1985) assumed that wetted soil volume depends upon the hydraulic conductivity of the soil, discharge of the emitter and amount of water available in the soil. They developed the following empirical equations to estimate the wetted depth and width. The equations were derived using three-dimensional cylindrical flow geometry and results were verified from plane flow model.

$$Z = 29.2(V_w)^{0.63}\left(\frac{K}{q}\right)^{0.45} \qquad \qquad ...(2.6)$$

$$w = 0.031(V_w)^{0.22}\left(\frac{K}{q}\right)^{-0.17} \qquad \qquad ...(2.7)$$

By combining the above two equations, we can find out the relationship between depth of wetting front, Z and width of wetted soil volume (w). The relationship can be expressed as follows:

$$w = 0.0094(Z)^{0.35}\, q^{0.33} K^{-0.33} \qquad \qquad ...(2.8)$$

where

$\quad Z$ = depth of wetting front in m

$\quad w$ = wetted width or diameter of wetted soil in m

$\quad V_w$ = volume of water applied in L

K = saturated hydraulic conductivity of soil, m/s

q = discharge of emitter, lph

Example 2.1. In a banana orchard, emitters of 4 lph discharge capacity are operating. The soil is sandy loam and rooting depth is 1.2 m. Saturated hydraulic conductivity of the soil is 30 mm/h. Find the width of the wetted soil volume.

Sol.: It is given:

$q = 4$ lph

Rooting depth is 1.2 m. It will be taken as vertical depth of wetting front. So, it is Z.

$K = 30$ mm/h $= 8.33 \times 10^{-6}$ m/s

The equation for width of wetted soil volume is

$$w = 0.0094 (Z)^{0.35} \, q^{0.33} K^{-0.33}$$

or $\qquad w = 0.0094 (1.2)^{0.35} \, 4^{0.33} \left(8.33 \times 10^{-6} \right)^{-0.33}$

$$= 0.75 \text{ m}$$

Ans. 0.75 m

2.5.1 Percentage Wetted Area

The percentage wetted area is defined as the average horizontal area wetted in the top 30 cm of the crop root zone as a percentage of the total crop area. For single lateral line, the percentage wetted area is computed from the following formula:

$$P_w = \frac{N_p \times S_e \times w}{S_p \times S_r} \qquad\qquad ...(2.9)$$

where,

P_w = percentage wetted soil area along a horizontal plane 30 cm below the soil surface, %

S_e = spacing between the emitters along a lateral line

N_p = number of emitters per plant

$S_p \times S_r$ = plant to plant and row to row spacing, m × m.

Example 2.2. A field is covered with cabbage and one emitter is supplying water to four plants. Cabbage is grown at the spacing of 60 × 60 cm. Find the percentage wetted for single line of lateral. Wetted width of soil volume is 0.75 m.

Sol.: It is given that $N_p = 1/4$, $S_e = 60$ cm $= 0.60$ m, $w = 0.75$ m

Now, the formula will be applied.

$$P_w = \frac{N_p \times S_e \times w}{S_p \times S_r}$$

$$\frac{\frac{1}{4} \times 0.60 \times 0.75}{0.60 \times 0.60} = 31.25\% \text{ or } 32\%$$

Ans. 32%

Mohammed (2010) developed a simple empirical model to determine the wetting pattern geometry from surface point-source drip irrigation system. The wetted soil volume was assumed to depend on the saturated hydraulic conductivity, volume of water applied, average change of moisture content and the emitter application rate. The following assumptions were made.

- A single surface point source irrigated a bare soil with a constant discharge rate.
- The soil is homogeneous and isotropic.
- No water table present in the vicinity of root zone.
- The evaporation losses are negligible.
- The effect of soil properties is represented by its porosity and saturated hydraulic conductivity.
- The value of porosity equals the value of saturated moisture content. It could be obtained using an equation given by Hillel (1982) which states:

$$n = \theta_S = \left(1 - \frac{\rho_b}{\rho_p} \right) \qquad ...(2.10)$$

where,

n = porosity of the soil

θ_s = Moisture content at 0 bars

ρ_b = bulk density of the soil (measured)

ρ_p = particle density of the soil

They considered that wetted radius and wetted depth of soil volume depends upon certain variables. The functional relationship among all the variables can be defined as follows:

$$r \propto f_1 (K, n, q_w, V_w) \qquad ...(2.11)$$

$$z \propto f_2 (K, n, q_w, V_w) \qquad ...(2.12)$$

where,

r = wetted radius

K = soil hydraulic conductivity

n = soil porosity

q_w = application rate

V_w = volume of water applied

z = depth of wetted zone

If we consider the equation given by Hillel (1982), the above two equations can be written as

$$r \propto f_1\left(K, \theta_S, q_w, V_w\right) \qquad \qquad ...(2.13)$$

$$z \propto f_2\left(K, \theta_S, q_w, V_w\right) \qquad \qquad ...(2.14)$$

Ben-Asher et al. (1986) investigated the infiltration from a point drip source in the presence of water extraction using an approximate hemispherical model. For infiltration from a point source without water extraction, they established the following:

$$\Delta\theta \approx \frac{\theta_S}{2} \qquad \qquad ...(2.15)$$

The new variable $\Delta\theta$ is called the average change of soil moisture content. This leads to:

$$r \propto f_1\left(K, \Delta\theta, q_w, V_w\right) \qquad \qquad ...(2.16)$$

$$z \propto f_2\left(K, \Delta\theta, q_w, V_w\right) \qquad \qquad ...(2.17)$$

According to the approaches introduced by Shwartzman and Zur (1986) and Ben Asher et al. (1986), the non-linear expressions describing wetting pattern may take the general forms as:

$$r = \Delta\theta^{\alpha} V_w{}^{\beta} q_w{}^{\gamma} K^{\lambda} \qquad \qquad ...(2.18)$$

$$z = \Delta\theta^{\rho} V_w{}^{\sigma} q_w{}^{\delta} K^{\varsigma} \qquad \qquad ...(2.19)$$

Once they identified the model structure and order, the coefficients were estimated in some manner. To determine the coefficients of Eqns. 2.18 and 2.19, four available published experimental data of Taghavi et al. (1984) Anglelakis et al. (1993), Hammami et al. (2002), and Li et al. (2003) were adopted. The choice of these experiments was essentially based on availability of their convenient data. A non-linear regression approach was used to find the best-fit parameters for the Eqns. 2.18 and 2.19. The following equations are obtained:

$$r = \Delta\theta^{-0.5626} V_w^{\,0.2686} q_w^{\,-0.0028} K^{-0.0344} \qquad \qquad ...(2.20)$$

$$z = \Delta\theta^{-0.383} V_w^{\,0.365} q_w^{\,-0.101} K^{0.1954} \qquad \qquad ...(2.21)$$

where, r and z (cm) are consistent units used in this approximations, V_w (ml), q_w (*ml/h*), and K in (cm/h).

Cook et al. (2006) developed a model and implemented in the WetUp software which uses data on the approximate radial and vertical wetting distances for different soils and discharge rates estimated by using analytical methods. WetUp is an easy to use and freely available software tool (http://www.clw.csiro.au/products/wetup), which will definitely help to graduate students to fine tune their research work on crop water management under drip irrigation systems. The program is a result of collaborative efforts among the Commonwealth Scientific and Industrial Research Organization (CSIRO), Cooperative Research Centre (CRC) for Sustainable Sugar Production and the National Program for Irrigation Research and Development (NPIRD) in Australia and the methods described by Thorburn et al. (2002). You cannot provide the manual inputs but can always select all the required inputs from pre-defined selection boxes and drop down menus.

Every simulation window opens with predefined values and the user can easily adjust or select the soil type, emitter flow rate, the maximum time and whether a surface or buried emitter should be simulated starting under dry, moist or wet soil condition. Different soils can be chosen by double clicking on the 'Select Soil Type' section on the simulation windows, or by choosing an appropriate button in the button bar. There are currently 29 soils which are based on average soil properties published by Clapp and Hornberger (1978) and measured field soils from Queensland in Australia published by Verburg et al. (2001). Flow rates may be selected in the range of 0.5 to 2.7 L/hr for an irrigation time of 1 to 24 hours. The depth of buried drip lines may be changed in the range of 0.1 to 1.5 m. In Indian conditions, we normally use 4 to 8 lph emitters. Since only 29 soils have been included, you may not find your soils and hence this is a limitation. However, suppose you want to simulate wetting pattern of 4 lph emitter after 12 hrs, then you can think of using emitter of 2 lph and simulate it for 24 hrs.

This software is meant as an educational tool and you can at least see the effect of changing the variables on wetting patterns.

3

COMPONENTS OF DRIP IRRIGATION SYSTEMS

A drip irrigation system consists of the components such as pump unit, fertigation equipment, filters, main, sub-main, laterals, and distributary outlets (emitters, microsprinklers, bubblers, etc.). Here we are concerned about drip system so we will talk about emitters in the following discussion. Besides, gate valves, check valves, pressure gauges and flow control valves are also used to regulate the flow of water and serve as additional components. Several major components of drip irrigation system are shown in Fig. 3.1. The components of drip irrigation system can be grouped into two major heads as:

(i) Control head

(ii) Distribution network

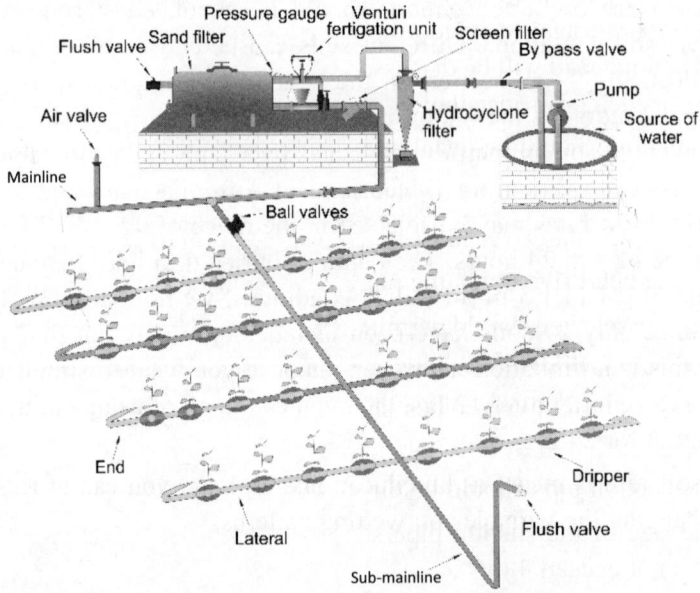

Fig. 3.1: Typical layout and the components of drip irrigation system

3.1 Control Head

The control head of drip irrigation system includes the pump or overhead tank, fertigation equipment, filters, and pressure regulator.

3.1.1 Pump/Overhead Tank

It is required to provide sufficient pressure in the drip irrigation system. Centrifugal pumps are generally used for low pressure drip irrigation systems. Overhead tank is generally used for small areas of orchard crops with a comparatively less water requirement.

The drip irrigation system requires energy to move water through the distribution pipe network and discharge it through emitters. In most irrigation systems, energy is imparted to water by a pump that in turn receives its energy from either an electric motor or an internal combustion engine. Therefore, it is important that both the pump and the engine be well suited to satisfy the requirements of the irrigation system. Usually, centrifugal pumps are used for this purpose. The characteristic curves of the pump are considered in selection of pumps. The characteristic curves show the relationship between capacity, head, power and efficiency of the pump. The head-capacity curve will give discharge of a pump at a given head. As the discharge increases, the head decreases. The pump efficiency increases with an increase in discharge but after a certain discharge, efficiency decreases. The brake horsepower (BHP) curve for a centrifugal pump increases over most of the range as the discharge increases. The pump horsepower at a maximum efficiency would be determined from the characteristic curves based on the irrigation system design discharge and the total dynamic head against which the pump is to operate. The total discharge and total dynamic head will be discussed later in this book. The following points should be considered for installation of a centrifugal pump.

- The pump should be installed as close to the water source as possible.
- Foundations should be rigid enough to absorb all vibrations.
- The pump and driver must be aligned carefully.
- On the belt drive units, the pump and driver shaft must be parallel.
- The site selected should permit the use of minimum possible connections on suction and delivery pipes.
- Suction and delivery pipes should be supported independently of the pump.
- The suction pipe should be direct and short.
- The size of the suction pipe should be such that the velocity of water does not exceed 3 m/s.

3.1.2 Fertigation

Fertigation means to apply fertilizers with the irrigation water. Chemical fertilizers are applied whenever needed by the crop in the appropriate form and quantity. The promotion of efficient, and effective water and fertilizer use is identified as an important contribution to the strategy needed to address problems of water scarcity and practicing intensive agriculture. Improving the water and consequently fertilizer use efficiency at farmers level, is the major contributor to increase food production and reverse the degradation of the environment or avoid irreversible environmental damage and allow for sustainable irrigated agriculture. Fertigation was proposed as a means to increase efficient use of water and fertilizers, increase yield, protect environment and sustain irrigated agriculture. Achieving maximum fertigation efficiency requires knowledge of crop nutrient requirements, soil nutrient supply, fertilizer injection technology, irrigation scheduling, and crop and soil monitoring techniques.

Fertigation is directly related with improved irrigation systems and water management. Drip and other microirrigation systems, highly efficient for water application, are ideally suited for fertigation. Water-soluble fertilizers at concentrations required by crops are conveyed through the irrigation systems to the wetted volume of soil. Thus, the distribution of chemicals in the irrigation water will likely place these chemicals in the root zone which helps the plants to uptake of N, P and K effectively. Many studies have been undertaken nationally and internationally to determine the effect of fertigation method on growth and yield of several fruits and vegetable crops.

Shedeed et al. (2009) evaluated the effect of method and rate of fertilizer application under drip irrigation system on growth, yield and nutrient uptake by tomato grown on sandy soil. The experiment was laid out in a randomized complete block design having five treatments replicated three times in 4.5 m × 12 m plot and included: control, normal fertilizers applied to soil with furrow irrigation, normal fertilizers applied to soil with drip irrigation, ½ soil – ½ fertigation, ¼ soil – ¾ fertigation and 100% NPK fertigation as water soluble fertilizers applied through drip fertigation. They concluded that the fertigation at 100% NPK recorded significantly higher total dry matter (4.85 t/ha) and leaf area index of 3.65, respectively, over normal fertilizer applied to soil with drip irrigation. It was also observed that the drip fertigation had the potential to minimize leaching loss and to improve the available K status in the root zone for efficient use by the crop. Drip fertigation helped in alleviating the problem of K deficiency in the sandy soil. Frequent supplementation of nutrients with irrigation water increased the availability of N, P and K in the root zone, which in turn influenced the yield and quality of tomato.

3.1.2.1 Advantages

The salient advantages of fertigation are given below.

- Uniform distribution of the nutrients in the soil by the irrigation water.
- Deep penetration of the nutrients into the soil.
- Lower fertilizer losses from the soil surface.
- Better coordination of nutrient supply with the changing crop nutrient requirements during the growing cycle.
- Higher application efficiency.
- Full control and precise dosing of fertilizers in automatic and semi-automatic irrigation systems avoid the leaching of nutrients beneath the root zone.
- Avoiding of mechanized fertilizer broadcast eliminates soil compaction and damage to the plants and the produce.
- Fertigation contributes to labor saving and convenience in fertilizer application.
- Improves soil solution conditions.
- Flexibility in timing of fertilizer application.

3.1.2.2 Limitations and Precautions

The limitations and precautions of using chemicals/fertilizers with drip irrigation are given below.

- Only completely soluble fertilizers are suitable to be applied through the irrigation system.
- Acid fertilizers can corrode metallic components of the irrigation system.
- The immersed fertilizers can raise the water pH and trigger the precipitation of insoluble salts that will clog the emitters and the filtration system.
- Operation and maintenance requires skilled manpower.
- Health hazard exists if the irrigation water system is connected to the drinking water supply network. Failure in the water supply may cause the back flow of water containing fertilizers into the drinking water system.
- The operator is exposed to burn injury by acid fertilizer solutions.
- To avoid the corrosion of metal parts, the pure water should be used in the last minutes of each irrigation cycle to wash the irrigation system from residual chemicals.

3.1.2.3 Fertigation Methods

The drip irrigation system employs injection devices for injecting the fertilizer into irrigation water. The injection device must ensure supply of constant concentration of fertilizer solution during the entire irrigation period. The common methods of applying chemicals/fertilizers through the drip irrigation system are pressure differential, injection pump, and venturi appliance as shown in Figs 3.2 to 3.4. In pressure differential systems as shown in Fig. 3.2, a pressure drop is created by pressure reducing valves provided between the inlet and the outlet of the supply tank. The pressure difference causes water flow through the tank and then chemicals/fertilizers from the tank are carried to the drip system. The tank contains solid soluble or liquid fertilizer that is dissolved gradually in the flowing water. The disadvantage of this type of injection system is that the concentration of fertilizer is diluted as injection continues. There are no moving parts and hence prolong its working life. The system is simple in operation and involves no extra cost.

In injection pump method, the fertilizer solution is injected by means of an injection pump (Fig. 3.3). A pump can be driven by electric motor, diesel engine, or hydraulically operated by the water pressure of the irrigation system. The hydraulic pump is versatile, reliable and has low operation and maintenance expenses. A pump must develop a pressure greater than that of in irrigation pipeline. Centrifugal pumps are used when high capacity injection is needed. Fertigation pumps can be controlled by the automatic irrigation system. The discharge of the pump is monitored by means of a pulse transmitter that is mounted on the pump and converts its piston or diaphragm oscillation into electrical signals. This system is flexible and can provide higher discharge rate and also does not create any further head loss in the irrigation system. It maintains a constant concentration of fertilizer solution throughout the application time. High cost of pump and its accessories, and operation and maintenance cost are its major disadvantages.

In venturi appliance, the fertilizer is injected through a constricted water flow path. A venturi injector is a tapered constriction which operates on the principle that a pressure drop results from the change in velocity of the water as it passes through the constriction. The velocity of water flow in the constricted section is increased. The pressure drop through a venturi must be sufficient to create a vacuum relative to atmospheric pressure in order to suck the solution from a tank into the injector. A tube mounted in the constricted section sucks fertilizer solution from an open fertilizer tank. A venturi injector does not require external power to operate. There are no moving parts, which increase their life and decreases probability of failure. The injector is usually constructed of plastic, which makes it resistant to most chemicals. It requires minimal operator attention and maintenance, and its cost is low as compared to other equipment of similar function and capability.

It is easy to adapt to most irrigation systems, provided a sufficient pressure differential can be created to suck the fertilizers. It is possible to inject nutrients in non-continuous (bulk) or continuous (concentration) fashion. For bulk injection, drip irrigation systems should be brought up to operating pressure before injecting any fertilizer or chemical. Fertilizer should be injected in a period such that enough time remains to permit complete flushing of the system without over-irrigation.

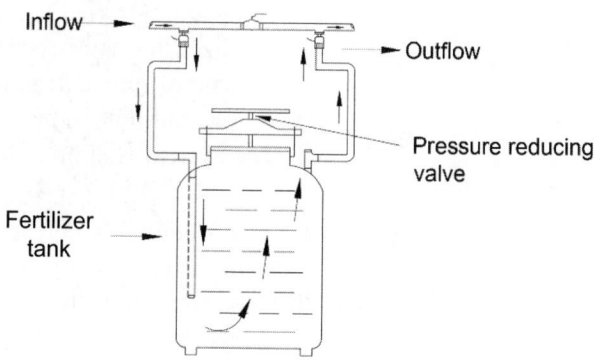

Fig. 3.2: Pressure differential system in closed fertilizer tank

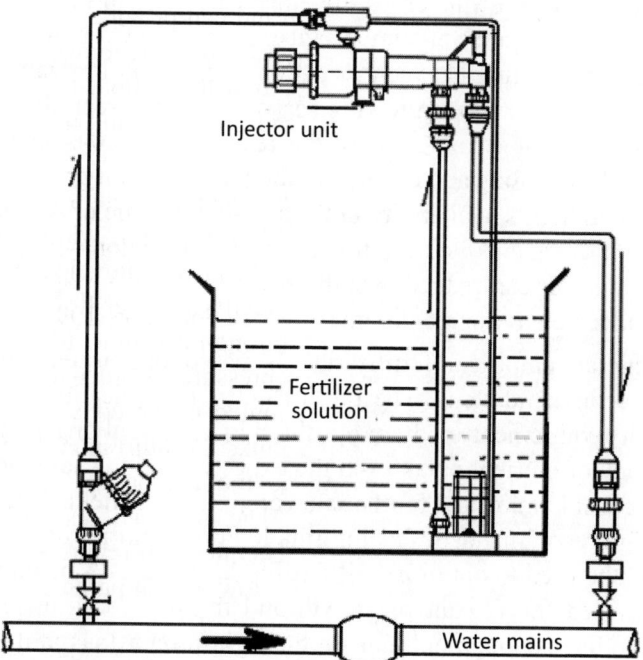

Fig. 3.3: Fertilizer injection methods by pump injection

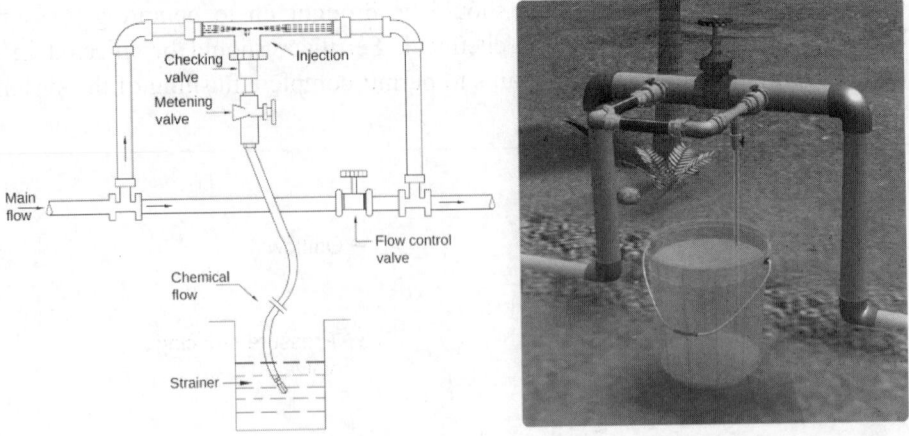

Fig. 3.4: Fertilizer injection methods by venturi

3.1.2.4 Fertigation Recommendations

For optimum plant growth and yield performance under fertigation, all fertilization-irrigation-input factors must be kept in mind so that none impose a significant limit. Implementing a fertigation program, the actual water and nutrient requirements of the crops, together with a uniform distribution of both water and nutrients, are very important parameters. Crop water requirements are the most critical link between irrigation and a good fertigation. The amount of irrigation water for entire growing season must be precisely estimated under the prevailing climatic conditions of the region under consideration. The main elements for formulating and evaluating the fertigation program are crop nutrient requirement, nutrients availability in the soil, the volume of soil occupied by the crop rooting system and the irrigation method. Fertigation with drip irrigation, if properly managed, can reduce overall fertilizer and water application rates and minimize adverse environmental impact (Papadopoulos, 1993).

In general, empirical fertilization is based on farmer's experience and on broad recommendations. The rate of nutrient application is determined by the nutrient requirement of the crop, the nutrient supplying power of the soil, the efficiency of nutrient uptake and the expected yield. These factors should be taken into consideration and for the same crop, for each field, different fertigation programs are recommended. The information on quantities of nutrients removed by crop can be used to optimize soil fertility level. Part of the nutrients removed by crop is used for vegetative growth and the rest for fruit production. It is important to have enough nutrients in the right proportions in the soil to supply crop needs during the entire growing season. Vegetable crops differ widely in their macronutrient requirements and in the pattern of uptake over the growing season. In general, N, P, and K uptake follow the same course as the rate of crop biomass accumulation. Fruiting crops such as tomato, capsicum, melon, etc. require relatively little nutrition until flowering and nutrient uptake accelerates and reaches to peak during fruit set and early fruit bulking.

Fertilization recommendations, based on research conducted regionally, vary among areas of the country. It is important to recognize these regional differences. Most of the agricultural universities in India have their own regional research station or Krishi Vigyan Kendra. The relevant research findings of these research stations must be utilized while formulating fertigation programs. Further, using a standard drip fertigation program without soil testing will often lead to wasteful fertilizer application or may result in a nutrient deficiency. A soil test helps to estimate the nutrient supplying power of a soil and reduces guesswork in fertilizer practices. Soil testing laboratories normally suggest ways to collect the soil samples. We are concerned here with drip irrigation systems so the sampling must be done accordingly. The place and depth of soil sampling relative to the drippers is a sensitive issue of particular importance.

Usually, it is recommended to get samples beneath the dripper, between the drippers and between the lateral pipes. In order to estimate the nutrient supplying capacity of a soil, apart from soil analysis, the parameters such as depth of the crop rooting system, percentage of soil occupied by the root system under different irrigation systems, and soil bulk density are needed. These parameters are used to calculate the weight of soil of a certain area to a depth where the active rooting zone of the crop is developed and estimate the reserve available nutrients for the crop. The appearance, growth and depth to which roots penetrate in soils are in part species properties.

For drip irrigated greenhouse vegetables like tomato, cucumber, and capsicum, the wetted soil volume is usually 30–50% of total soil volume. The fraction of soil occupied by roots must be taken into account whenever the amount of available nutrients is calculated; otherwise the available amounts could be overestimated. In calculating the nutrient supplying capacity of a soil, the whole amount of the available nutrient to full depletion of soil can be taken into consideration. However, it is preferable that a certain amount of a nutrient be reserved in soil. For intensive irrigated agriculture as safety amounts of P and K in soil could be considered the 30 and 100 ppm, respectively. Moreover, in case that a nutrient is below the safety value, the fertilization program may include an amount of nutrient needed to build up soil fertility up to the safety margin. These margins are at the same time the pool for increased demand in nutrients at eventual crop critical nutrient stages. It should be emphasized that the amount of fertilizer nutrients needed by the crop and the amount of nutrients, which should be applied, are not equivalent. The crop does not use all the nutrients supplied by fertilizers, therefore, the actual amount applied is higher than the amount required by the crop. In general, the higher the water use efficiency of an irrigation system, the higher is the nutrient uptake efficiency. For a well-designed drip irrigation system and with good scheduling of irrigation, depending on soil type, the potential N, P and K uptake efficiency ranges between 0.75–0.85, 0.25–0.35 and 0.80–0.90, respectively.

The capacity of the injection system depends upon the concentration, rate and frequency of application of fertilizer solution. The amount of fertilizer to be applied per application (P) can be calculated by

$$P = \frac{F \times A}{N} \qquad\qquad ...(3.1)$$

where,

F = fertilizer requirement (kg/ha)

A = area of the field (ha)

N = number of applications

The injection rate depends upon the concentration of the liquid fertilizer and desired quantity of nutrients to be applied during the irrigation. It can be determined by the equation.

$$q_c = \frac{F \times A}{c \times t_r \times T} \qquad\qquad ...(3.2)$$

where,

q_c = rate of injection of liquid fertilizer solution into the system (lph)

F = fertilizer application rate per irrigation cycle (kg/ha)

A = area irrigated (ha)

T = irrigation application time (h)

c = concentration of nutrients in liquid fertilizer (kg/L)

t_r = ratio between fertilizing time and irrigation application time

Fertigation rate and time depend on the irrigation water flow rate and fertilizer application rate. For most cases, 0.1% of the irrigation flow rate is a proper injection rate (Burt et al., 1995). Few steps are needed, if we want to determine the injection rate and time for fertigation, if a liquid fertilizer is used. First step is to calculate the total amount of nitrogen needed by multiplying the number of plants or by farm size in hectares. Next step is to find out the total weight of fertilizer needed for fertigation. The total weight depends on both total N to be applied and the grade of the selected nitrogen fertilizer. The standard crop-nutrient rating (NPK rating) of urea is 46–0–0. Hence, it contains 46% elemental nitrogen (N), 0% elemental phosphorous (P), and 0% elemental potassium (K). Urea contains 46% N by weight. The total weight of the fertilizer to apply is equal to the total N needed divided by the N concentration (0.46 in this case). Finally, the injection rate is determined by dividing the irrigation water flow rate by the dilution factor [(0.46 × 1,000,000)/100 ppm]. Injection time is determined by dividing the number of litres of liquid N fertilizer needed by the injection rate.

Let us do it with a simple example of Banana plantation (2 × 2m) in one hectare of land where we will use a liquid fertilizer CAN-46 (not a commercial name) with 46% N and density of 1.32 kg/L. Nitrogen requirement per plant is 250 g. Irrigation flow rate is 10 lps and concentration of N in irrigation line is to be maintained at 150 ppm.

Step 1: Total nitrogen $= 250 \text{ g/plant} \times \dfrac{10000}{2 \times 2} = 625000 \text{ g}$

$= 625 \text{ kg}$

Step 2: Total weight of CAN-46 $= \dfrac{625}{0.46} = 1360 \text{ kg}$

Step 3: Litres of CAN-46 $= \dfrac{1360}{1.32} = 1030 \text{ Lit}$

Step 4: Dilution factor $= \dfrac{0.46 \times 1000000}{150} = 3066$

Step 5: Injection rate $= \dfrac{10 \times 3600}{3066} = 11.74 \text{ lph}$

Step 6: Injection time $= \dfrac{1030}{11.74} \approx 88 \text{ hr}$

Therefore, in this particular case, 1030 L of CAN-46 are needed for the complete fertigation event. This can be divided along with number of irrigation events. In that case, the fertilizer requirement of banana as 250 g needs to be divided. The fertilizer distribution efficiency of the system is determined by measuring the weight of the fertilizer (mg) in the total volume of water cached from the different emitter, during operation 30 min from the start of fertigation. The fertilizer distribution efficiency can be determined as follows:

$$FDU = \frac{W_{fm}}{W_{fa}} \qquad \qquad ...(3.3)$$

where,

FDU = fertilizer distribution efficiency (%)

W_{fm} = mean weight of fertilizer in water of the lowest 1/4 emitter (mg)

W_{fa} = mean weight of fertilizer in water during fertigation (mg)

3.1.2.5 *Plant Nutrients*

The nutritional elements of plant are divided into two groups:

(i) **Macroelements:** This group includes elements that are consumed by plants in relatively high amounts. The common macro elements are nitrogen, phosphorous, potassium, calcium, magnesium and sulfur.

(ii) **Microelements:** This group includes elements that are also essential for good development of plants, but they are absorbed and used by plants relatively in small quantity. Common microelements are iron, manganese, zinc, copper, molybdenum and boron.

The functions of plant nutrients which support the plant system are briefly discussed below.

Nitrogen (N): The nitrate form of nitrogen is not held in soils. Nitrates move with other soluble salts to the wetted front. This is of particular interest since NO_3-N should always be applied with irrigation and at desired concentration needed by the crop to satisfy its nitrogen requirement from one irrigation event to the other. Under irregular NO_3-N application, the fertigated crops might be under the overfertilization stage at the day of fertilizer application and under deficient stage due to leaching following the irrigation without fertilizer. The ammonium form of N derived from ammonium or urea fertilizers does not leach immediately because it is temporarily fixed on exchange sites in the soil.

Ammonium and urea, however, may induce acidification, which may create higher solubility and movement of phosphrous in the soil. Urea is a highly soluble, chargeless molecule, which easily moves with the irrigation water and is distributed in the soil similarly to NO_3. At 25°C, it is hydrolyzed by soil microbial enzymes into NH_4 within a few days.

Phosphorus (P): Phosphorus is essential for cell division and plant root development. It enhances flowering at the reproductive stage and is essential for development of seeds and fruits and necessary for meristematic division. Contrary to N and K, phosphorus is readily fixed in most of the soils. Movement of P differs with the form of fertilizer, soil texture, soil pH and the pH of the fertilizer. Phosphorus mobility in soil is very restricted due to its strong retention by soil oxides and clay minerals. Soil application of commonly available P fertilizers generally results in poor utilization efficiency mainly because phosphate ions rapidly undergo precipitation and adsorption reactions in the soil, which remove them from the soil solution.

Consequently, there is little or no movement of phosphate from point of contact with the soil. Therefore, there is inefficient utilization of applied P fertilizers. Rauschkolb et al. (1976) found that P movement increases 5–10 folds when applied through drip system, indicating that fertigation of P is particularly important.

Potassium (K): The ionic form of this element is K^+. Potassium takes part in activating enzymes involved in photosynthesis and in the metabolism for creating proteins and carbohydrates. The carbohydrates are transported from leaves to the roots and anions from roots to the leaves. It also assists in utilizing the water use by regulating the stomata and decreasing evaporation from the stomata. It improves the quality of fruits and vegetables. It is less mobile than nitrate, and distribution in the wetted volume may be more uniform due to interaction with soil binding sites.

Drip irrigation systems apply K in both laterally and downward direction, allowing more uniform spreading of the K in the wetted volume of soil. Application of K with the irrigation water is advised since its effectiveness increases substantially and boost higher crop yield. Potassium can be applied as potassium sulphate, potassium chloride and potassium nitrate. These potassium sources are soluble with little precipitation problems.

Calcium (Ca): The ionic form of calcium element is Ca^{2+}. This element is involved in cell structure by creating calcium pectate and the cell division. Calcium takes part in activating the reactions of some enzymes such as phospholipase, ATPase, etc. It reacts as a detoxifying agent.

Magnesium (Mg): The ionic form of magnesium element absorbed by the plants is Mg^{2+}. Magnesium is the major element in chlorophyll structure, which is responsible for photosynthesis process. It takes part in activating some enzymes that are involved in carbohydrates synthesis. It enhances the uptake and translocation of phosphate.

Sulfur (S): The ionic form of this element absorbed by the plant is SO_4^{2-}. Sulfur is a component in the structure of some amino acids such as cystine, cysteine and is active in protein structure. It is involved in some enzyme oxidation and reduction reactions.

Iron (Fe): The ionic forms of this element absorbed by the plants are Fe^{2+} and Fe^{3+}. This element is an important component in the chlorophyll synthesis in the plant. It is involved in enzymatic oxidation – reduction reactions.

Manganese (Mn): It is an activator in some enzymatic oxidation – reduction reactions. It takes part in activating the enzymes that are responsible for reduction of nitrate NO_3^- to nitrite NO_2^-. It is involved in chlorophyll synthesis and photosynthesis reactions.

Zinc (Zn): Zinc is involved in biosynthesis of indole acetic acid (IAA). It is involved in enzymatic reactions.

Copper (Cu): Copper is involved in some enzymatic reduction reactions like ascorbic acid oxidase, phenolase, lactase, etc.

Molybdenum (Mo): The ionic form of this nutritional element is MoO_4^{2-}. This activates enzymes that are responsible for the reduction of nitrate NO_3^- to nitrite NO_2^-, therefore in molybdenum deficiency nitrate is accumulated in plant tissues. It is required by the rhizobium bacteria for nitrogen fixation in legume.

Boron (B): The ionic form of this nutritional element is $B_4O_7^{2-}$. Boron activates certain enzymes and participates with sugars in producing complex compounds (lignine) that are responsible for the thickening and hardening of cell walls. It participates in the process of producing of rhizobium nodules that are responsible for the nitrogen fixation of legume roots.

The fertilizer to be used for fertigation must be water-soluble (Table 3.1). However, most of the common P and K fertilizers are not convenient for fertigation due to their low solubility. This is particularly the case with the P fertilizers.

Table 3.1: Solubility of fertilizers in water (kg/100 liters)

Type of fertilizer	*Solubility*
Ammonium sulphate	71
Ammonium nitrate	119
Urea	110
Monoammonium phosphate (MAP)	23
Urea phosphate	96
Potassium sulphate	7
Potassium nitrate	32

Example 3.1. The nitrogen concentration of solid urea fertilizer is 46%. Calculate the amount of urea to prepare a solution of 80 ppm nitrogen. Consider specific weight of urea as 1.32 g/cm^3.

Sol.: The amount of urea to be dissolved in 1 m^3 of water can be calculated by

$$\frac{\text{ppm of element} \times 100}{\% \text{ of element in fertilizer}}$$

$$= \frac{80 \times 100}{46 \times 1.32} = 131.75 \text{ grams or } 131.75 \text{ mg/L}$$

Ans. 131.75 mg/L

Example 3.2. The liquid ammonium nitrate has nitrogen concentration of 18% with specific weight of 1.23 g/cm^3. Determine the amount of liquid ammonium nitrate to prepare a solution of 100 ppm nitrogen.

Sol.: The amount of liquid ammonium nitrate to be dissolved in 1 m^3 of water can be calculated by

$$\frac{\text{ppm of element} \times 100}{\% \text{ of element in fertilizer} \times \text{Specific weight}}$$

$$\frac{100 \times 100}{18 \times 1.23} = 451.67 \text{ ml}$$

Ans. 451.67 ml

Example 3.3. A drip irrigation system is installed in 1 ha area under citrus crop. The plants are spaced at 5 m × 5.5 m apart. Urea–ammonium nitrate, a liquid fertilizer with 32% nitrogen and weighing 1.32 kg/L is available for fertigation. Find the fertilizer injection rate to apply 40 kg/ha of elemental nitrogen. The irrigation and fertilizing time are 8 hrs and 4 hrs, respectively.

Sol.:

Irrigation application time $(T) = 8$ h

Fertilizing time $= 4$ h,

The ratio of fertilizing time and irrigation application time $(t_r) = 4/8 = 0.5$

The concentration of elemental nitrogen (N) in the liquid fertilizer

$c = 1.32 \times 32/100 = 0.42$ kg/L

Using Eqn. 3.2, the injection rate

$$q_c = \frac{40 \times 1.0}{0.42 \times 0.50 \times 8} = 23.80 \text{ lph}$$

Ans. 23.80 lph

3.1.2.6 Sources of Fertilizers

Commercial fertilizers may also precipitate in the irrigation pipe networks and react with ions present in the irrigation water such as Ca^{2+} or Mg^{2+}. Therefore, when choosing the P fertilizer for fertigation, besides solubility, we must take care to avoid P-Ca and P-Mg precipitation in the irrigation lines and emitters. Keeping this in view, acid P fertilizers like phosphoric acid, urea phosphate or monoammonium phosphate are recommended. Different sources of fertilizers, including P fertilizers, have different effects on irrigation water and soil pH. High pH values greater than 7.5 in the irrigation water are undesirable. Calcium and Mg carbonate and orthophosphate precipitations may occur in the lines and the drippers. In addition, high pH may reduce Zn, Fe and P availability to plants.

The desired pH is below 7 and most cultivated crops are grown in the range of 5.5–6.5. The pH of the irrigation water could be reduced or controlled by using P acid or acid based fertilizers like urea phosphate and monoammonium phosphate. The use of acid fertilizers in drip systems may be beneficial in many ways other than the direct benefit from the added P, such as increased solubility of soil native P minerals, increased availability of other nutrients and micronutrients and prevention of clogging of the fertigation system.

The choice of fertilizer suitable for a specific application should be based on several factors such as nutrient form, purity, solubility, and cost. A variety of fertilizers can be injected into drip irrigation systems. Soluble NPK fertilizers are available in the market which are appropriate for fertigation but the price might be in certain cases the main constraint. Common N sources include ammonium sulphate, urea, ammonium nitrate, urea-ammonium nitrate, calcium nitrate, magnesium nitrate and potassium nitrate. Potassium can be supplied from potassium chloride, potassium sulfate, potassium thiosulfate, or potassium nitrate. In case that salinity is a problem, potassium chloride and potassium sulphate should be avoided. For phosphorus application, phosphoric acid, urea phosphate or ammonium phosphate solutions are used commonly. Mono-ammonium or mono potassium phosphate is also available.

3.1.2.7 *Nutrient Monitoring*

It is essential to monitor the soil and plant nutrient status to ensure maximum crop productivity. In conventional production, soil NO_3-N testing is usually carried out before we plant. Since drip irrigation system provides the ability to add N as and when required, more extensive NO_3-N monitoring is justified. Traditional soil laboratory test offers the most complete and accurate information. There are several alternative techniques to aid on-farm nitrogen measurement. One approach is the use of soil solution access tubes, also called suction lysimeters. The details are not being discussed here. The use of suction lysimetry has serious limitations. There can be large spatial variability; one portion of a field may vary from another and, since NO_3-N moves with the wetting front, there can be stratification of NO_3-N within the bed. This problem can be minimized by using multiple lysimeters per field, but that also increases the effort and the cost. Interpretation of results is also problematic.

In general, when NO_3-N concentration in a root zone soil solution is found greater than 75 mg/L, it indicates that sufficient N is available to meet immediate plant needs. A lower NO_3-N concentration cannot be interpreted directly as N deficiency, given the difficulty to obtain a sample representative of the whole root zone. Reliance on soil NO_3-N testing is appropriate early in the crop cycle, when crop N uptake rate is low and the detection of substantial residual NO_3-N can lead to reduced additional N fertigation. By mid-season, crop uptake rates increase and soil NO_3-N concentration correspondingly will change more rapidly. Also, once an extensive root system is developed, many crops can take up N in excess of crop needs. From mid-season until harvest, plant tissue analysis should be the primary indicator of N status, although soil testing still may be used to identify fields where NO_3-N levels remain high enough to delay additional N application.

Conventional plant tissue analysis, in which tissue is dried, grounded and analyzed chemically in a laboratory, is the most accurate way to determine crop nutrient status. Unfortunately, laboratory analysis of dry tissue is relatively costly, and the time lag between sampling and obtaining results can be significant. In recent years, there has been increasing interest in on-farm tissue testing, particularly for monitoring drip-irrigated fields. On-farm monitoring usually involves the analysis of NO_3-N and K content of petiole sap; sap analysis for PO_4^--P is uncommon. Measurement techniques include colorimetric methods, NO_3-N or K test strips (Hochmuth, 1994), or ion-specific electrode (Hartz et al.,

1993; Vitosh and Silva, 1994). Although all methods can be used successfully, the ion-specific electrode is the most commonly used approach. The appropriate protocol for tissue collection, handling, and analysis is discussed by Hochmuth (1994).

3.1.3 Filtration System

The clogging of emitters is the main problem encountered in the operation of drip irrigation systems. Filtering and keeping contaminants out of the system are the main defense against the clogging caused by mineral and organic particles. Impurities in water can be classified into three categories:

(*a*) Inorganic solid particles: Sand, silt and clay and insoluble precipitation.

(*b*) Living organisms, such as algae, protozoa, bacteria, and fungi.

(*c*) Organic debris.

Removal of above mentioned impurities is essential for efficient and trouble-free operation of a drip irrigation system which necessitates the use of proper filters.

3.1.3.1 Selection of filters

While selecting filters for the drip irrigation system, the following factors are considered:

- The physical quality of water such as the concentration and pattern of the impurities, suspended solids and organic matter.
- The chemical nature of water such as pH level, and presence of sediments forming chemical elements and possible reaction with the injected fertilizers when fertigation is applied.
- Discharge and allowable head losses in the system.
- Reliability and durability of the filters.
- Cost of the filter.
- The total surface area of the filtration element is very important. The filtration area needed for moderately dirt water is in the range of 60–150 cm^2 for drip irrigation.

3.1.3.2 Types of Filters

Settling Basins: Settling basins can remove suspended material ranging from sand (2000 μm) to silt (200 μm) in stream water being used for irrigation. It removes large volumes of sand and silt. Basins are constructed so that it could limit turbulence and permit a minimum of 15 minutes of retention time for water to travel from the basin inlet to the pumping system intake. Longer retention time is required to allow the settling of smaller particles. A basin of 1.2 m deep, 3.3 m wide and 13.7 m long is required to provide a one-quarter hour retention time for a 57 lps stream. Settling basin should be relatively long and narrow to eliminate short circuit current that reduces effective retention time. If the source of water is groundwater, settling basins should not be used, but if canal water is being used for irrigation through drip irrigation systems, it may be of a great use.

Media Filter: Media filters are used when irrigating with water containing high organic load such as water pumped from open water bodies or reclaimed water. It consists of fine gravel and sand of selected sizes placed in a pressurized tank. Arrangement of laying gravels/sand is shown in Fig. 3.5. Media filters are not easily plugged by algae and can remove relatively large amounts of suspended solids before cleaning is needed. It can retain particle sizes in the range of 25 to 200 μm. In general water flow rates through the filters should be in between 10 lps and 18 lps per square meter of filtration surface area.

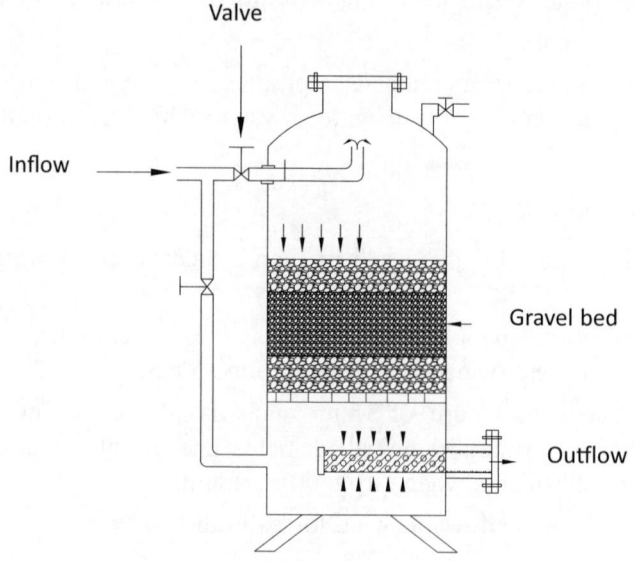

Fig. 3.5: Arrangements of gravels in media filter

Media filter should be followed by a secondary screen filter to prevent carry-over of contaminants following the backwashing process. Numbers designate the sand media filter used in most drip irrigation filters: Number 8 and 11 are crushed granite, and numbers 16, 20, and 30 are silica sands. The mean granule size in microns for each media number is approximately 1900, 1000, 825, 550 and 340 for number 8, 11, 16, 20, and 30, respectively.

When the water flows from top of the filter, it goes down through layered bed and flows out at the bottom of the filter. The organic impurities adhere to the surface of the media particles and accumulate in the filter tank. When too much dirt has been accumulated in the layered bed, the pressure head loss across the filter is increased and the accumulated dirt has to be backwashed in time in order to avoid the excess head losses. Backwashing is a process of reversing the direction of water flow through the filter and bypassing the effluent and its need can be detected by the pressure drop across the filter. The American Society of Agricultural Engineers recommends that the pressure drop in the media filter should not exceed 70 kPa. They do not remove very fine silt and clay particles as well as bacteria.

Centrifugal Filters: The best treatment of the water containing soil particles is sedimentation of the particles by means of sand separators. Sand separators, hydrocyclones or centrifugal filters remove suspended particles that have specific gravity greater than 1.2. The centrifugal sand separator separates the sand and other heavy particles from the water by means of centrifugal force of the tangentially entering water into a conic tank (Fig. 3.6).

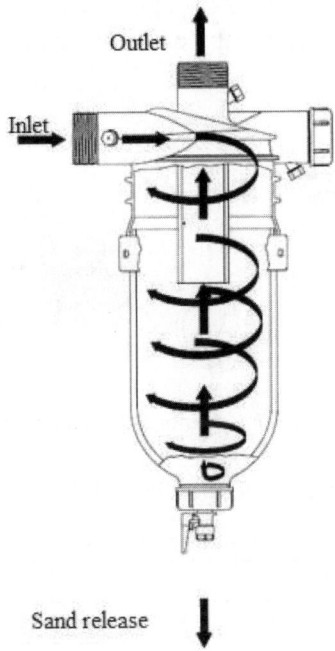

Fig. 3.6: Centrifugal filter

The sand is thrown by the centrifugal force against the conic wall, settled down and accumulated in a collector at the bottom. The collector is washed out manually when full. The clean water is emitted in a spiral motion through an outlet at the top of the separator. The diameters of the top and bottom of the conical shape in centrifugal filter are designed in accordance with the flow of water. The centrifugal filters effectively remove large quantity of sand particles. They are placed often at the upstream of media or screen filters.

They are ideal for situations where a pile of sand is present in water. A sand separator filter (hydrocyclone) is used in combination with a filter media after it has an excellent combination. It should be noted that selection of hydrocyclone sand separator filter must be closely combined with the flow rate of the system. A minimum pressure head-loss is necessary to achieve effective sand separation.

Screen Filter: Screen filters are fitted just after the pumping unit and no other filter is required if source of water is groundwater. The casing is built of metal or plastic material. It has four apertures: Water inlet, outlet, draining valve and cover. It consists of a screen made of metal, plastic, or synthetic cloth enclosed in a special house used to limit maximum particle size. Screens are classified according to the number of openings per inch with standard wire size for each screen size (Table 3.2). Most manufacturers recommend 100 to 200 mesh screens for drip irrigation system. Normally, the discharge through the screens is less than 135 lps per square meter of screen openings. A standard 200 mesh stainless steel screen has only 58% open area and equivalent nylon mesh with same size opening has only 24% open area. In screen filter, the mesh size and the total open area determine the efficiency and operational limits.

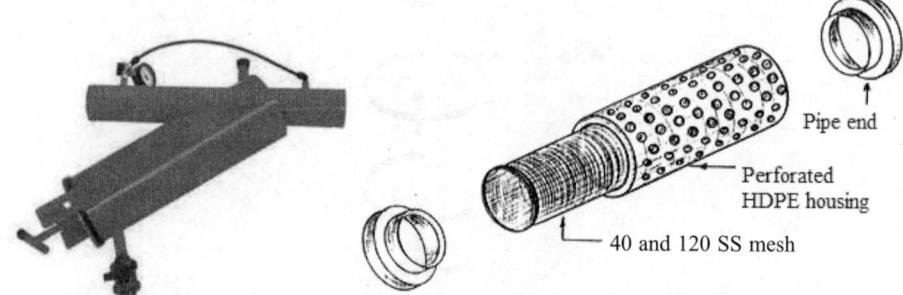

Fig. 3.7: Schematic views of components of screen filter (Jain irrigation systems)

Table 3.2: Specifications of filter screens

Mesh No.	Hole size (*micron*)	Wire thickness (*micron*)
40	420	250
50	300	188
80	177	119
100	149	102
120	125	86
155	100	66
200	74	53

Therefore, it is important to consider the percentage of open area when sizing a filter for a given system discharge. It may be fitted in a series with the gravel filter in order to remove the solid impurities like fine sand, dust, etc. from water. Screen filters are suitable to water with inorganic impurities. High load of organic and biological impurities may clog the screen promptly. Like media filters, the screen filters can be cleaned manually or automatically (Fig. 3.7).

Disk Filters: This is more suitable for water with mixed impurities of inorganic solid particles and organic debris. The casing is made of metal or plastic material. The filtration element is made of stacked grooved ring shaped disks, tightened together by a threaded cap with a total equivalent screen size ranging from 40 to 400 mesh. Water is filtered as it flows through the grooves. Coarse particles are trapped on the external surface of the stack while finer particles and organic debris adhere to the inner grooves. The discs are pressed together during filtration and direction of flow is reversed during back flushing mode. The discs start a spinning motion and complete retained impurities are removed. The discs are assembled in a cylinder and are pressed together. To handle large flow rate, they can be installed in batteries. The disk filters have much more higher dirt retention capacity than screen filters with the same specifications. Such filters are often used as final filtering component before the water enters the system (Fig. 3.8).

Fig. 3.8: Disc filters

Example 3.4. How many 1.22 m diameter sand filters are needed for a 50 ha drip irrigated farm? Crop ET is 10 mm/day. Irrigation application efficiency is 90%. Maximum flow rate through sand filters should be 18 lps per square meter of tank cross-sectional area.

Sol.: If the crop evapotranspiration requirement is 10 mm/day, the approximate gross irrigation requirement will be 10/0.90 or 11.11 ≈ mm/day = 0.011 m/day.

$$Q = 50 \times 10{,}000 \times 0.011 = 5500 \text{ m}^3$$

Area of the cross section of the tank $= \frac{\pi}{4}D^2 = \frac{\pi}{4}(1.22)^2 = 1.17 \text{ m}^2$

Total discharge required to be passed through filter

$$= \frac{5500 \times 1000}{24 \times 3600} = 63.65 \approx 64 \text{ lps}$$

Number of filters $= \frac{64 \times 1.17}{18} = 4.16 \approx 5$

Ans. 5

3.1.3.3 Filter Characteristics and Evaluation

The field evaluation of drip irrigation system includes a determination of filtration efficiency and pressure differential across the filter. The effectiveness of the filtration system can also be indirectly assessed by evaluating the degree of emitter plugging. The filtration efficiency (F_r) of a filtration system is estimated by the following formula:

$$F_r = \left(1 - \frac{S_{out}}{S_{in}}\right) \times 100 \qquad \qquad ...(3.4)$$

where,

    S_{out} = the concentration of suspended solids (mg/L) in filter outlet, and

    S_{in} = the concentration of suspended solids (mg/L) in filter inlet.

The water sample should be taken at least 30 minutes after the system is turned on. The second aspect of field evaluation of filters in a drip irrigation system is differential pressure before and after filter backwashing takes place. If the differential pressure is very high, a loss of flow can occur which affects the downstream pressure. Another approach for determining the effectiveness of the system is the extent of emitter plugging. A small percentage of emitter plugging can affect the uniformity of water application. Pressure regulators are generally used to decrease the higher system pressure to the lower required system pressure. It controls the pressure in one way only, i.e. high to low.

3.2 Distribution Network

The distribution network constitutes mainline, sub-mainline and laterals with drippers and other accessories.

3.2.1 Main and Sub-mainline

A typical drip mainline is generally made of rigid Polyvinylchloride (PVC) and high density polyethylene (HDPE). Pipes of 63 mm diameter and above with a pressure rating 4 to 6 kg/cm^2 are generally used for mainline pipes. The sub-main pipeline is made of rigid PVC, HDPE or LDPE (low density polyethylene) of outside diameter ranging from 50 to 75 mm with a pressure rating of 2.5 kg/cm^2. The diameters of main and sub-main pipelines are chosen based on estimated water to be carried out which ultimately need the careful determination of peak crop water requirement. The sub-mailline is connected to mainline to deliver the water to laterals. Gate valve is provided on the sub-mail line to control the flow and practiced when irrigating large area divided in many sub-plots. Each sub-plot will be irrigated with one sub-main. Sub-mainline is fitted with a flushing valve to remove the contaminants. The main and sub-mainlines are usually placed underground.

3.2.2 Laterals

Laterals are the pipes normally manufactured from LDPE or linear low density polyethylene (LLDPE). Generally, pipes having 12, 16 and 20 mm internal diameter with a wall thickness varying from 1 to 3 mm are used as laterals. Nowadays, even 32 mm lateral pipe is also being used. Since laterals are always laid over the surface, it must be flexible and non-corrosive. Pressure variation within the lateral lines must be within acceptable limits and therefore, selection of length and diameter of lateral line is a matter of accurate design.

3.2.3 Drippers / Emitters

The emitters are connected to the laterals and control the flow of water coming out of laterals. The design of single outlet emitter is based on the principle of energy losses by sudden change in velocity of fluid flow in the emitter path. The successive change in velocity occurs due to sudden enlargement and contraction in the designed flow path of the disc element. The necessary arrangement for expansion of flow is done by making larger cross-sectional area in the shape of a circle and contraction of flow by smaller cross-sectional area in the shape of a rectangular channel. The bottom of the path is made flat to increase the wetted perimeter of the flow passage, which leads to decrease the hydraulic radius and velocity of flow in the flow passage (Fig. 3.9). The energy losses occur when the flow channel suddenly expands to a larger diameter circular path.

The energy loss is accomplished by two ways—first due to the impulse momentum which takes place as the water flows from narrower rectangular passage to the wider circular passage and secondly when the water flows from wider circular section to the narrower channel section. In this process, a lot of eddies are formed which causes a considerable dissipation of energy. The high pressure from the lateral line transmitted into the emitter is greatly reduced and controlled flow of water emits in discrete drops almost at atmospheric pressure. Emitters are made from polypropylene or polyethylene and available in the market in different types and designs.

Point-source and line-source emitters operate either above or below the ground surface. Most of the point-source emitters are either on-line or in-line emitters. The primary difference between on-line and in-line emitters is that the entire flow required downstream of the emitter passes through an in-line emitter. There is more head loss along a lateral with on-line emitters than one with in-line emitters because of obstruction created by the barbs of on-line emitters. The percentage area wetted, and the reliability of the emitters against the clogging and malfunctioning are two important aspects of quality and safety of drip irrigation systems.

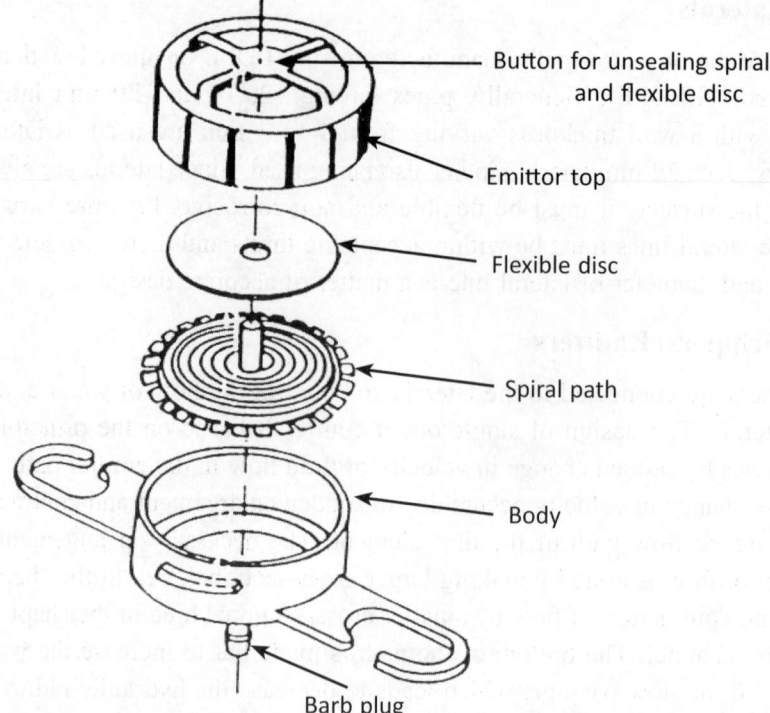

Fig. 3.9: Components of a point-source on-line pressure compensating dripper

An ideal set of emitters should have the attributes such as durability, low cost, reliable performance with a relatively low rate of uniform discharge and relatively large and/or self-flushing passage way to reduce or prevent clogging.

3.2.3.1 Types of Drippers/Emitters

(a) **In-line Emitters**: In-line emitters are fixed along with the lateral line. The pipe is cut and dripper is fixed in between the cut ends, such that it makes a continuous row after fixing the dripper. They have generally a simple thread type or labyrinth type flow path (Fig. 3.10).

(b) **On-line Emitters:** These are fixed on the lateral by punching suitable size holes in the pipe. These are of the following types:

  (i) **Simple Type/Laminar Flow:** In this type of dripper, the discharge is directly proportional to the pressure. They have simple thread type, labyrinth type, zig-zag path, vortex type flow path or have float type arrangement to dissipate energy.

  (ii) **Turbo Key Drippers:** These are made of virgin and stabilized polymers and are available in 2, 4 and 8 lph discharge. They provide resistance to blockage and are pressure compensating.

Fig. 3.10: In-line dripper

(iii) **Pressure Compensating Drippers:** This type of dripper gives a fairly uniform discharge within the pressure range of 0.3 atmospheres to 3.5 atmospheres. They are provided with a high quality rubber diaphragm to control pressure and are most suitable on slopes and difficult terrain.

(iv) **Built-in Dripper Tube:** In this system, polyethylene drippers are inseparably welded to the inside of the tube during extrusion of polyethylene pipes. They are provided with independent pressure compensating water discharge mechanism and extremely wide water passage to prevent clogging. Other accessories include take out/ starter, rubber grommet, end plug, joints, tees and manifolds.

3.2.3.2 *Emitter Discharge Exponent*

The flow from the emitter used in drip irrigation system is expressed by the following relationship:

$$q = K_d H^x \qquad \qquad ...(3.5)$$

where, q is discharge rate of emitter in lph, K_d is discharge coefficient which is a constant of proportionality, H is the pressure head at emitter in meter, and x is emitter discharge exponent. The value of x governs the flow regime, and discharge and pressure relationship of an emitter. The pressure variation will less affect the discharge with lower value of x. Normally, x is taken as 0.5 for non-compensating simple orifice and nozzle emitters. For fully compensating emitter, $x = 0.0$. The exponent of long path emitter is normally varies from 0.7 to 0.8. The values of x usually lie between 0.5 and 0.7 for tortuous path emitters.

3.2.3.3 Selection of an Emission Device

The selection of an emission device such as line-source and point-source emitters, microsprinklers and bubblers depends upon crop to be irrigated, filtration requirement and soil type. The required emitter flow rate can be computed as follows:

$$Q_r = \frac{d_i \times I_i}{I_t \times E_a \times N} \qquad \qquad ...(3.6)$$

where,

Q_r = required emitter flow rate (L/h)

d_i = water requirement of per plant (L/d)

I_i = irrigation interval (day)

I_t = irrigation time per set (h)

E_a = application efficiency (fraction)

N = number of emitters per plant

The crop water requirement under drip irrigation is different from the crop water requirement under surface and sprinkler irrigation primarily because land area wetted is reduced resulting in less evaporation from soil surface. The capacity of emission device can be computed by using following equation:

$$Q = \frac{A \times d \times 100}{(H - T_m) \times E_a} \qquad \qquad ...(3.7)$$

where,

Q = capacity of emission device (L/h)

A = area irrigated by the emission device (m²)

d = depth of applied water (mm)

H = hours of application of irrigation water

T_m = off time for maintenance (h)

E_a = application efficiency (%)

The area irrigated by an emission device is computed by the following equation:

$$A = \frac{L \times S \times W_p}{100 \times N_e} \qquad \qquad ...(3.8)$$

where,

A = area irrigated (m²)

L = spacing between adjacent plant rows (m)

S = spacing between emission point (m)

W_p = % of cropped area being irrigated

N_e = number of emission devices at each emission point

The value of W_p varies from 30 to 100%, depending upon the type of crop and its age. For widely spaced horticultural fruit tress (vine, bush, mango, etc.), W_p varies between 30 to 60 % of area of each tree and for close growing crops such as vegetables, the value of W_p is considered as 100%. The number of emission devices per emission point, N_e required for desired wetting pattern is determined on the basis of horizontal and vertical movement of water through the soil. The soil moisture movement studies are required to be conducted at several representative sites around the field to obtain horizontal and vertical spread data.

3.2.3.4 Emitter Spacing

As we have discussed in Chapter 2 that wetting pattern of the emitter determines the spacing and number of emitter required per plant. To determine the wetted width and depth of water front, Mohammed (2010), Schwartzman and Zur (1986) and Wetup model can be used. United States Soil Conservation Services (1984) has provided an estimate of maximum horizontal wetted width from a single outlet and this can be used where filed data are not available (Table 3.3). For line-source emitter, this value must be multiplied by 0.8.

Table 3.3: Wetted width or diameter of wetted circle of a point-source emitter of 4 lph capacity (United States Soil Conservation Services, 1984)

Depth of root zone and soil texture	Homogeneous soil layer (m)	Soil profiles of varying textural group	
		Low density (m)	Moderate density (m)
Root zone depth, 0.75 m			
Coarse soil	0.45	0.75	1.05
Medium soil	0.90	1.20	1.50
Fine soil	1.05	1.50	1.80
Root zone depth, 1.5 m			
Coarse soil	0.75	1.40	1.80
Medium soil	1.20	2.10	2.70
Fine soil	1.50	2.00	2.40

4

ESTIMATION OF
CROP WATER REQUIREMENT

The quantity of water used by the crop is directly proportional to its growth. Crops use water in transpiration. The root hairs extract water from the soil and the water travels through the stem towards the leaves. The water transpires into the air through pores in the surface of the leaves. Water is also lost when it evaporates from the soil and other surfaces. The combined loss of water through transpiration and evaporation is termed as evapotranspiration. Measurement of the evapotranspiration will tell you how much water you are required to supply to the crop.

The amount of water used by the crop will depend upon the type of crop and its stage of growth, soil types, and environmental factors such as sunlight, humidity, wind speed and temperature. The crops grown in areas which are hot, dry, windy and sunny will require more water than crops grown in cool, humid and cloudy environment. The influence of the climate on crop water requirement is accommodated by the reference evapotranspiration (ET_0). It can be defined as the rate of evapotranspiration from a large area, covered by green grass, 8 to 15 cm tall, which grows actively, completely shades the ground and which is not short of water and expressed in mm per unit of time, e.g. mm/day or mm/month. Grass has been taken as the reference crop.

4.1 Methods of Evapotranspiration Estimation

4.1.1 Direct Methods of Measurement of Evapotranspiration

4.1.1.1 Pan Evaporation Method

Evaporation pans provide a measurement of the combined effect of temperature, humidity, wind speed and sunshine on the reference crop evapotranspiration. Many different types of evaporation pans are being used. The best known pans are the Class A evaporation pan (circular pan) as shown in Fig. 4.1 and the Sunken Colorado pan (square pan). The evaporation pan is installed in the field or at the experimental site and filled with a known quantity of water. The water depth in the pan is to be recorded and the water is allowed to evaporate for 24 hours. After 24 hours, the remaining quantity of water or say water depth will be measured. If the area receives any rainfall, that may be recorded.

If the water depth in the pan drops too much, water is added and the water depth is measured before and after the water is added. If the water level rises too much, water is taken out of the pan and the water depths before and after is measured. The amount of evaporation (E_{PAN}) per time unit is calculated. This alone cannot be considered as the water requirement of the crops grown nearby areas. The E_{PAN} is multiplied by a pan coefficient, K_{PAN}, to obtain the reference evapotranspiration (ET_0).

$$ET_0 = K_{PAN} \times E_{PAN} \qquad \qquad ...(4.1)$$

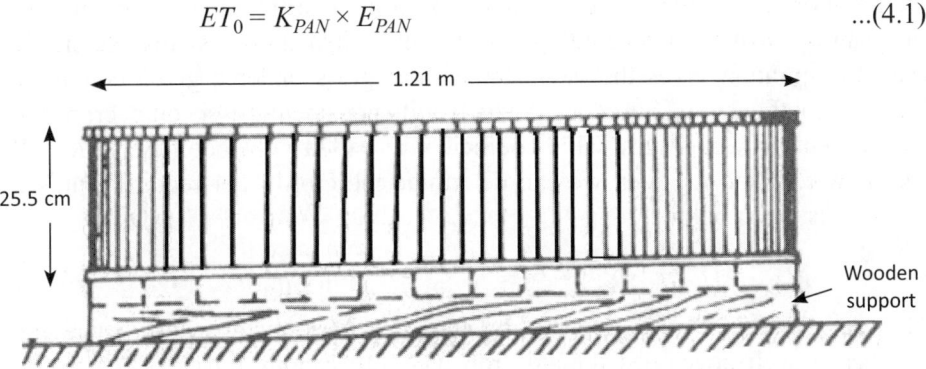

Fig 4.1: Class A evaporation pan

Determination of K_{PAN}

It is obvious that the water in the evaporation pan and from the grass field does not evaporate in a similar fashion under the same climate. Therefore, a special coefficient is used (K_{PAN}) to relate one to the other. The pan coefficient, K_{PAN}, depends on the type of pan used, the pan environment, and the climate (humidity and wind speed). For a pan placed in a fallow area with high humidity and low wind speed, the K_{PAN} will be high. But if the pan is placed in a cropped area with low humidity and higher wind speed, the K_{PAN} will be low. Normally, for the Class A evaporation pan, the K_{PAN} varies between 0.35 and 0.85. Average K_{PAN} may be taken as 0.70. For the Sunken Colorado pan, the K_{PAN} ranges between 0.45 and 1.10, and average K_{PAN} is normally taken as 0.80.

Example 4.1. Class A evaporation pan is installed in a field and first day the water depth was recorded as 60 mm and after 24 hours water depth becomes 45 mm. The day received no rainfall. If K_{PAN} is 0.70, find out the reference evapotranspiration.

Sol.: We know the formula to estimate reference evapotranspiration which is given hereunder.

$$ET_0 = K_{PAN} \times E_{PAN}$$

Pan evaporation = 60 − 45 = 15 mm

$$ET_0 = 0.70 \times 15$$

$$= 10.5 \text{ mm/day}$$

Ans. 10.5 mm/day

Crop Coefficient

We are more interested in the estimation of crop water requirement so that we can plan our irrigation scheduling and design of irrigation systems. Reference evapotranspiration gives the value for a short grass under a given condition. Here, a coefficient is introduced which will consider the specific crop and its different growth stages. This coefficient is called crop coefficient (K_c). If we know ET_0 and K_c, then we can determine the crop water requirement, i.e. ET_{CROP}, as given below.

$$ET_{CROP} = K_c \times ET_0 \qquad \qquad ...(4.2)$$

The crop coefficient, K_c, depends on the type of crop, growing stages and weather. A well developed tomato crop with full canopy cover will transpire more water than the reference grass crop. The same crop during initial stages may require less water than the reference grass crop. I mean to say that the value of the crop coefficient ranges from the initial to final harvesting stages. Also in order to determine the K_c value for any crop, knowledge about the length of the crop growing season and the lengths of the various growth stages is essential. Food and Agriculture Organization (FAO) has given all these information but it is better to collect the total length of growing season and the duration of the various growth stages of the crops from nearby agricultural university or agricultural research station. The total growing season has been divided into 4 growth stages: Initial stage, crop development stage, mid-season stage, and the late season stage.

Table 4.3 shows the total length of growing periods and duration of the various growth stages for some of the major field crops (FAO). Table 4.4 shows the average K_c values for some crops according to its four different growth stages. Actually, the K_c value also depends upon the weather, especially on the relative humidity and the wind speed. The values indicated in Table 4.4 should be reduced by 0.05, if the relative humidity is high (RH > 80%) and the wind speed is low (u < 2 m/sec). The values may be increased by 0.05 for relative humidity lesser than 50% and the wind speed more than 5 m/sec.

Now, the values of crop coefficient can be taken from the table given by FAO but growth stages are not confined on monthly basis. For example, crop development stage may range from January 15 to February 20. How will you calculate the monthly crop coefficient? We are going to explain with an example.

Annual mean and monthly mean of precipitation, maximum and minimum temperature, potential evapotranspiration and reference evapotranspiration can be obtained from www.indiawaterportal.org. Average daily potential evapotranspiration and average daily reference evapotranspiration (mm/day) for Ranchi region is given in Tables 4.1 and 4.2. Methods of estimation of reference evapotranspiration will be discussed later in this chapter.

Table 4.1: Average daily potential evapotranspiration (mm/day) for Ranchi, Jharkhand

Year	Jan	Feb	Mar	Apr	May	Jun	Jul	Aug	Sep	Oct	Nov	Dec
2000	5.29	5.22	6.84	8.36	8.57	7.06	5.36	5.31	5.08	5.81	5.69	5.22
2001	5.23	5.83	6.93	8.15	8.66	6.69	5.05	5.22	5.35	5.50	5.40	4.97
2002	5.08	5.86	7.15	8.30	8.75	7.46	5.79	5.11	5.26	5.86	5.68	5.06

Table 4.2: Average daily reference evapotranspiration (mm/day) for Ranchi, Jharkhand

Year	Jan	Feb	Mar	Apr	May	Jun	Jul	Aug	Sep	Oct	Nov	Dec
2000	3.01	3.46	5.06	6.73	6.98	5.83	4.41	4.28	3.92	4.02	3.5	2.93
2001	3	3.82	5.25	6.49	7.14	5.52	4.22	4.26	4.14	3.91	3.32	2.98
2002	3.05	3.93	5.41	6.62	7.32	6.16	4.71	4.2	4.13	3.94	3.37	3

Example 4.2. Tomatoes are transplanted on 1st November and normally these are harvested up to March. The monthly evapotranspiration are given and average relative humidity is around 80%. The prevalent wind speed is 4 m/s. Calculate the total crop water requirement of tomatoes.

Months	November	December	January	February	March
ET_0 (mm/day)	5.0	4.5	4.0	6.0	6.5

Sol.: Let us first determine the duration of various growth stages as follows.

Crops	Total growing season	Initial growth stage	Crop development stage	Mid-season stage	Late season stage
Tomatoes	145	30	40	50	25

We know that the total duration of crop is 145 days and date of transplanting of tomatoes is 1st November, now total duration of growing period can be written datewise in the following manner. K_c values have been taken from Table 4.4 keeping in view that humidity and wind speed are normal.

Growth stages	Date	K_c values
Initial stage, 30 days	1 Nov – 30 Nov	0.45
Crop development stage, 40 days	1 Dec – 9 Jan	0.75
Mid-season stage, 50 days	10 Jan – 28 Feb	1.15
Late season stage, 25 days	1 Mar – 25 Mar	0.80
Last date of harvest	25 Mar	---

Now, you see that months and growing stages are not matching, i.e. crop development stage is covering December and 9 days in January and we have K_c values according to the crop development stages. This has to be corrected in the following manner.

November : $K_c = 0.45$

December : $K_c = 0.75$

January : $K_c = \frac{9}{30} \times 0.75 + \frac{22}{30} \times 1.15 = 1.07$

February : $K_c = 1.15$

March : $K_c = 0.80$

Now we have monthly evapotranspiration and K_c values, we can determine crop water requirement on monthly basis. All months are assumed to have 30 days as suggested by FAO.

$$ET_{CROP} = K_c \times ET_0$$

November : $ET_{CROP} = 5 \times 0.45 \times 30 = 67.5$ mm/month

December : $ET_{CROP} = 4.5 \times 0.75 \times 30 = 101.1$ mm/month

January : $ET_{CROP} = 4 \times 1.07 \times 30 = 128.4$ mm/month

February : $ET_{CROP} = 6 \times 1.15 \times 30 = 207.0$ mm/month

March : $ET_{CROP} = 6.5 \times 0.80 \times 30 = 130.0$ mm/month

After adding the monthly crop water requirement, total water requirement for tomato is obtained as **634 mm**.

Ans. 634 mm

Table 4.3: Approximate duration of growth stages for various field crops

Field crops	*Total*	*Initial stage*	*Crop development stage*	*Mid-season stage*	*Late season stage*
Barley/Oats/ Wheat	120	15	25	50	30
	150	15	30	65	40
Bean (green)	75	15	25	25	10
	90	20	30	30	10
Bean (dry)	95	15	25	35	20
	110	20	30	40	20
Cabbage	120	20	25	60	15
	140	25	30	65	20
Carrot	100	20	30	30	20
	150	25	35	70	20
Cotton/Flax	180	30	50	55	45
	195	30	50	65	50
Cucumber	105	20	30	40	15
	130	25	35	50	20
Eggplant	130	30	40	40	20
	140	30	40	45	25
Grain (small)	150	20	30	60	40
	165	25	35	65	40
Lentil	150	20	30	60	40
	170	25	35	70	40
Lettuce	75	20	30	15	10
	140	35	50	45	10
Maize (sweet)	80	20	25	25	10
	110	20	30	50	10
Maize (grain)	125	20	35	40	30
	180	30	50	60	40
Melon	120	25	35	40	20
	160	30	45	65	20

Contd...

Field crops	Total	Initial stage	Crop development stage	Mid-season stage	Late season stage
Millet	105	15	25	40	25
	140	20	30	55	35
Onion (green)	70	25	30	10	5
	95	25	40	20	10
Onion (dry)	150	15	25	70	40
	210	20	35	110	45
Peanut/Groundnut	130	25	35	45	25
	140	30	40	45	25
Pea	90	15	25	35	15
	100	20	30	35	15
Pepper	120	25	35	40	20
	210	30	40	110	30
Potato	105	25	30	30	20
	145	30	35	50	30
Radish	35	5	10	15	5
	40	10	10	15	5
Sorghum	120	20	30	40	30
	130	20	35	45	30
Soybean	135	20	30	60	25
	150	20	30	70	30
Spinach	60	20	20	15	5
	100	20	30	40	10
Squash	95	20	30	30	15
	120	25	35	35	25
Sugarbeet	160	25	35	60	40
	230	45	65	80	40
Sunflower	125	20	35	45	25
	130	25	35	45	25
Tomato	135	30	40	40	25
	180	35	45	70	30

Table 4.4: Values of the crop factor (K_c) for various crops and growth stages

Crop	Initial stage	Crop dev. stage	Mid-season stage	Late season stage
Barley/Oats/Wheat	0.35	0.75	1.15	0.45
Bean (green)	0.35	0.70	1.10	0.90
Bean (dry)	0.35	0.70	1.10	0.30
Cabbage/Carrot	0.45	0.75	1.05	0.90
Cotton/Flax	0.45	0.75	1.15	0.75
Cucumber/Squash	0.45	0.70	0.90	0.75
Eggplant/Tomato	0.45	0.75	1.15	0.80
Grain/small	0.35	0.75	1.10	0.65
Lentil/Pulses	0.45	0.75	1.10	0.50
Lettuce/Spinach	0.45	0.60	1.00	0.90
Maize (sweet)	0.40	0.80	1.15	1.00
Maize (grain)	0.40	0.80	1.15	0.70
Melon	0.45	0.75	1.00	0.75
Millet	0.35	0.70	1.10	0.65
Onion (green)	0.50	0.70	1.00	1.00
Onion (dry)	0.50	0.75	1.05	0.85
Peanut/Groundnut	0.45	0.75	1.05	0.70
Pea (fresh)	0.45	0.80	1.15	1.05
Pepper (fresh)	0.35	0.70	1.05	0.90
Potato	0.45	0.75	1.15	0.85
Radish	0.45	0.60	0.90	0.90
Sorghum	0.35	0.75	1.10	0.65
Soybean	0.35	0.75	1.10	0.60
Sugarbeet	0.45	0.80	1.15	0.80
Sunflower	0.35	0.75	1.15	0.55
Tobacco	0.35	0.75	1.10	0.90

4.1.1.2 Soil Moisture Depletion Method

The soil moisture depletion technique is usually employed to determine the consumptive use of irrigated field crops grown on fairly uniform soil, where the depth of groundwater is such that it will not influence the soil moisture fluctuations within root zone. This technique involves the measurement of soil moisture of various depths in effective root zone at a number of times throughout the crop growth period. The greater the number of measurements, the more will be accuracy. By summing moisture depletion of each of the soil layer and for each interval for the growth period, ET can be determined as:

$$U = \sum_{i=1}^{n} \left(\frac{M_1 - M_2}{100} \right).B_i.D_i + ER \qquad ...(4.3)$$

where,

U = moisture depleted (cm) from the soil profile at a certain interval

M_1 = moisture content (%) in ith layer at the time of first sampling

M_2 = moisture content (%) in ith layer at the time of next sampling

B_i = bulk density of ith the layer, g/cm^3

D_i = depth of the soil layer (cm)

ER = effective rainfall (cm)

For periods between irrigation day and sampling day, soil moisture depletion is assumed to be equal to pan evaporation. Then water used for crop growth period is determined by adding moisture depleted (U_i) during all intervals:

$$ET_{CROP} = \sum_{i=1}^{n} U_i + E_{PANC} \qquad ...(4.4)$$

here, E_{PANC} is cumulative pan evaporation between irrigation and sampling day.

Evaporation can be measured by using a lysimeter which gives additional information on soil water balance. A lysimeter is a measuring device used to measure the amount of actual evapotranspiration. By recording the amount of precipitation and the amount lost through the soil, the amount of water lost to evapotranspiration can be calculated. Lysimeters are of two types: Weighing and non-weighing. We have not included the detail discussion on lysimeter.

4.1.2 Empirical Methods

4.1.2.1 Thornthwaite Method

This is a widely used method for estimating potential evapotranspiration, developed by Thornthwaite (1948) who correlated mean monthly temperature with evapotranspiration as determined from water balance for valleys where sufficient moisture water was available to maintain active transpiration. This formula is based mainly on temperature and estimates of the unadjusted potential evapotranspiration (mm), calculated on a monthly basis taking a standard month of 30 days and 12 hours of sunlight/day, is given by:

$$ET_0 = 16 \left(\frac{10\bar{T}_m}{I} \right)^a \qquad ...(4.5)$$

where, m is the months 1, 2, 3,..., 12, $\bar{T}_m$ is the mean monthly air temperature (°C); I is the heat index for the year, given by:

$$I = \Sigma i_m = \Sigma \left(\frac{\bar{T}_m}{5} \right)^{1.5} \quad \text{for } m = 1, 2, 3, \ldots, 12 \qquad \ldots(4.6)$$

a is an empirical exponent which can be computed by the following expression:

$$a = 6.7 \times 10^{-7} \times I^3 - 7.7 \times 10^{-5} \times I^2 + 1.8 \times 10^{-2} \times I + 0.49 \qquad \ldots(4.7)$$

Thornthwaite's equation was widely criticized for its empirical nature but also widely used. Because Thornthwaite's method of estimating ET_0 can be computed using only temperature, it has been one of the most misused empirical equations in arid and semi-arid irrigated areas where the requirement has not been maintained (Thornthwaite and Mather, 1955). The application of this method to compute evapotranspiration for short time periods usually leads to error because incoming radiations may not be reflected by short term temperatures.

Compared to the Penman formula, Thornthwaite values tend to exaggerate the potential evaporation. This is particularly marked in the summer months with the high temperatures having a dominant effect in the Thornthwaite computation, whereas the Penman estimate takes into consideration other meteorological factors.

4.1.2.2 Blaney-Criddle Method

Blaney and Criddle (1950) proposed an empirical method, a temperature-based approach for calculating potential evapotranspiration. The method uses temperature as well as day length duration, minimum daily relative humidity, and daytime wind speed at 2m height. The model is quite sensitive to the wind speed variable and somewhat insensitive to the estimate of relative humidity. The preferred form of the equation reveals its strong empiricism and is given by Shuttleworth (1993).

$$ET_0 = a + bf \qquad \ldots(4.8)$$

where,

$$f = p \, (0.46T + 8.13) \qquad \ldots(4.9)$$

$$a = 0.0004RH_{MN} - (n/N) - 1.4 \qquad \ldots(4.10)$$

$$b = 0.82 - 0.0041RH_{MIN} + 1.07(n/N) + 0.066(U_d)$$
$$- 0.006(RH_{MIN})(n/N) - 0.0006(RII_{MIN})(U_d) \qquad \ldots(4.11)$$

The variable p in Eqn. 4.9 is the actual daily daytime hours to annual mean daily daytime hours expresses as a percent, T is the mean air temperature in °C, (n/N) is the ratio of actual to possible sunshine hours, RH_{MIN} is the minimum daily relative humidity in percent, and U_d is the daytime wind at 2 m in m/s.

4.1.2.3 Hargreaves Method

The Hargreaves equation is a temperature based method and gives an expression for the reference crop evapotranspiration (Hargreaves and Samani, 1982,1985). It gives reasonable estimates of reference crop evapotranspiration because it takes into account the incoming solar radiation and the impact of radiation warming the surface. The Hargreaves equation takes the form;

$$ET_0 = 0.0022 \times R_A \times \delta T^{0.5} \times (T + 17.8) \qquad ...(4.12)$$

where,

R_A = mean extra-terrestrial radiation (mm/day)

δT = temperature difference of mean monthly maximum temperature and minimum temperature for the month (°C).

T = mean air temperature (°C).

4.1.3 Micrometeorological Method

4.1.3.1 Mass Transport Approach

As we know that the evaporation depends on the incoming heat energy and vapor pressure gradient, which, in turn, depend on several weather variables, such as temperature, wind speed, atmospheric pressure, solar radiation, and quality of water. The whole process of evaporation is complicated. A complete physical model cannot accommodate all the factors which are not controllable and measurable. The mass-transfer approach is one of the oldest methods (Dalton, 1802; Meyer, 1915; Penman, 1948) and is still an attractive method for estimating free water surface evaporation because of its simplicity and reasonable accuracy. The mass-transfer methods are based on the Dalton equation, which for free water surface can be written as:

$$E_0 = C\,(e_s - e_a) \qquad ...(4.13)$$

where, E_0 is free water surface evaporation, e_s is the saturation vapor pressure at the temperature of the water surface, e_a is the vapor pressure in the air and C is aerodynamic conductance. The term e_a is also equal to the saturation vapor pressure at the dew point temperature. The parameter C depends on the horizontal wind speed, surface roughness and thermally induced turbulence. It is normally assumed to be largely dependent on wind speed, u. Therefore, Eqn. 4.13 can be expressed as:

$$E_0 = f(u)\,(e_s - e_a) \qquad ...(4.14)$$

where, $f(u)$ is the wind function. This function depends on the observational heights of the wind speed and vapor pressure measurements. Penman (1948) proposed the following variation of the Dalton equation.

$$E_0 = 0.40\,(e_s - e_a)\,(1 + 0.17u_d) \qquad ...(4.15)$$

where,

E_0 = potential evaporation (mm)

e_s = saturation vapor pressure at the temperature of the water surface (mm Hg)

e_a = actual vapor pressure in the air (mm Hg)

u_d = wind speed at 2 m height (miles/day)

Example 4.3. Determine the potential evaporation with Penman equation. The following data set is given:

e_a = 11.50 mm Hg

e_s = 13.20 mm Hg

u_d = 45 miles/day

Sol.: The potential evaporation can be calculated by the following equation.

$$E_0 = 0.40 \ (e_s - e_a) \ (1 + 0.17 u_d)$$

$$= 0.40 \ (13.20 - 11.50) \ (1 + 0.17 \times 45)$$

$$= 0.68 \times 8.65$$

$$= 5.88 \ \text{mm/day}$$

Ans. 5.88 mm/day

Example 4.4. In a cropped field, five sets of the micrometeorological data were recorded and given in the table. Compute the potential evaporation by using the Penman mass transport method.

Data set	Saturation vapor pressure (mm Hg)	Actual vapor pressure (mm Hg)	Wind speed at 2 m height (miles/day)
1.	12.0	10.8	40.5
2.	10.2	8.2	12.8
3.	8.6	7.6	30.8
4.	7.2	7.0	45.6
5.	10.4	9.2	42.0

Sol.: Penman (1948) equation based on mass transport approach is given hereunder. All the required inputs are given.

$$E_0 = 0.40 \ (e_s - e_a) \ (1 + 0.17 u_d)$$

For data set 1.

Saturation vapor pressure at the temperature of the water surface, $e_s = 12.0$ mm Hg

Actual vapor pressure in the air, $e_a = 10.8$ mm Hg

Wind speed at 2 m height, $u_d = 40.5$ miles/day

Now, applying the formula, we get

$$E_0 = 0.40 \, (e_s - e_a) \, (1 + 0.17u_d)$$
$$= 0.40 \, (12.0 - 10.8) \, (1 + 0.17 \times 40.5)$$
$$= 3.78 \text{ mm/day}$$

Similarly for other data sets, the information is given in the table.

Data set	Saturation vapor pressure, e_s (mm Hg)	Actual vapor pressure, e_a (mm Hg)	Wind speed at 2 m height, u_d (miles/day)	Potential evaporation, E_0 (mm/day)
1.	12.0	10.8	40.5	3.78
2.	10.2	8.2	12.8	2.54
3.	8.6	7.6	30.8	2.49
4.	7.2	7.0	45.6	2.45
5.	10.4	9.2	42.0	3.91

4.1.3.2 Aerodynamic Approach

Evaporation from an open surface is a dynamic process, whose rate depends on ability of wind and the humidity gradient in the air above the evaporating surface to remove water vapor away from the evaporating surface via aerodynamic processes. Thornthwaite and Holzman (1939) proposed the following equation which included specific humidity gradient and logarithmic wind profile.

$$E_0 = \rho_a k^2 \frac{\left[(q_1 - q_2)(u_2 - u_1) \right]}{\left[\ln(z_2/z_1) \right]^2} \qquad \ldots(4.16)$$

where,

E_0 = evaporation (mm/hr)

ρ_a = density of air (kg/m^3)

k = von Karman's constant (0.41)

q_2 and q_1 = specific humidity in g/kg at heights z_2 and z_1

u_2 and u_1 = wind speed in m/s at heights z_2 and z_1

Example 4.5. Estimate the evaporation from field data collected in West Bengal by using aerodynamic approach.

The data are given below.

$$\rho_a = \text{density of air (kg/m}^3) = 1.3 \text{ kg/m}^3$$

$$k = \text{von Karman's constant, } 0.41$$

q_2 and q_1 = specific humidity in g/kg at heights z_2 and z_1 = 6.90 and 8.75 g/kg

$$u_1 = 1.42 \text{ m/s, } u_2 = 1.84 \text{ m/s, } z_2 = 2 \text{ m, } z_1 = 0.5 \text{ m}$$

Sol.: All the values are given required by the formula used in aerodynamic approach. It is simple and putting the given values in Eqn. 4.16,

$$E_0 = \rho_a k^2 \frac{\left[(q_1 - q_2)(u_2 - u_1)\right]}{\left[\ln(z_2/z_1)\right]^2}$$

$$= 1.3 \times 0.41^2 \frac{\left[(8.75 - 6.90)(1.84 - 1.42)\right]}{\left[\ln(2.0/0.50)\right]^2}$$

$$= 0.17/1.92 = 0.089 \text{ g/m}^2/\text{sec } (1 \text{ g/m}^2/\text{s} = 10^{-3} \text{ mm/s})$$

$$= (0.089/1000) \times 3600 = 0.32 \text{ mm/hr}$$

Ans. 0.32 mm/hr

4.1.4 Energy Balance Approach – Bowen Ratio Method

The Bowen ratio method is used because of the simplicity of data collection and robust nature of the system allowing for long-term data acquisition. The technique has increased in popularity because of the recent improvements in field portable data acquisition systems and sensor accuracy and precision. The technique was first proposed by Bowen (1926) which introduced a relationship between sensible heat flux (H) and latent heat (λE) known as Bowen ratio (β) and is given by:

$$\beta = \frac{H}{\lambda E} = \alpha \left(\frac{C_p \Delta T}{\lambda \Delta e}\right)$$

...(4.17)

where, H is sensible heat flux (W/m²), λE is latent heat flux (W/m²), α is the ratio of the turbulent transfer coefficients for sensible heat and water vapor (K_h / K_w), both in units of m²/s, C_p is specific heat of air at constant pressure (J/kg/°C), ΔT is the air temperature gradient (°C) between two heights above the surface, λ is latent heat of vaporization (J/kg), and Δe is the gradient of vapor pressure (kPa) at the same two heights as T under non-advective conditions. The surface energy balance can be expressed as:

$$R_n = G + H + \lambda E$$

...(4.18)

where, R_n is net irradiance (W/m^2), G is soil heat flux (W/m^2). Positive values of R_n and G are toward the surface, while H and λE are away from the soil surface. Using above two equations and solving for λE yields the following estimate of evaporation:

$$\lambda E = \frac{R_n - G}{1 + \beta}$$

...(4.19)

Limitations of the Bowen ratio method generally occur near sunrise and sunset, because of small gradients in T and e that result in p values approaching to -1 or ∞. Limitations can also occur with crops of non-uniform cover and under conditions of advection, commonly found in semiarid agriculture (Prueger et al., 1997).

Example 4.6. Field data were collected to estimate latent heat flux by using Bowen ratio method which are give below.

Net irradiance,	(R_n)	= 282 W/m^2
Soil heat flux,	(G)	= 10.1 W/m^2
Dry bulb temperature at 1 m height		= 17.0°C
Dry bulb temperature at 2 m height		= 16.8°C
Wet bulb temperature at 1 m height		= 9.8°C
Wet bulb temperature at 2 m height		= 8.7°C
Psychometric constant	γ	= 0.49 mm Hg/°C

Vapor pressure from hygrometric table at two heights, e_1 and e_2 are 1.55 and 0.96 mm Hg.

Sol.: The gradient of vapor pressure at 1 m and 2 m heights can be calculated as

$$\Delta e = e_1 - e_2$$
$$= 1.63 - 0.98$$
$$= 0.65 \text{ mm Hg}$$

Temperature gradient can be determined by taking the difference of dry bulb temperatures at two different heights, i.e.

$$\Delta T = 18.0 - 17.6$$
$$= 0.40°C$$

Now, since we have calculated all the factors needed to calculate the Bowen ratio (β), it can be determined as follows:

$$\beta = \alpha \left(\frac{C_p \Delta T}{\lambda \Delta e} \right) = 0.49 \left(\frac{1 \times 0.40}{1 \times 0.65} \right) = 0.30$$

$$= 0.49 \left(\frac{1 \times 0.4}{1 \times 0.65} \right)$$

$$= 0.30$$

Further, latent heat flux, λE, is determined with the help of Eqn. 4.19.

$$= \lambda E = \frac{R_n - G}{1 + \beta} = \frac{2.82 - 10.1}{1 + 0.30}$$

$$= 209.15 \times 4.12 \times 10^{-7} \times 3600 \text{ mm/hr}$$

$$= 0.310 \text{ mm/hr}$$

Ans. 0.310 mm/hr

Note: 1 W/m^2 = 4.12 × 10^{-7} mm/s

4.1.5 Combination Methods

Combination methods explain both the aerodynamic and energy balance approaches to estimate the evapotranspiration. These methods require sophisticated and expensive instruments for recording observation and not valid under moisture stress condition.

4.1.5.1 Penman Method

The energy balance at the earth's surface, which equates all incoming and outgoing energy fluxes can be expressed as follows.

$$R_n = H + \lambda E + G \qquad \qquad ...(4.20)$$

where,

R_n = energy flux density of net incoming radiation (W/m^2)

H = flux density of sensible heat into the air (W/m^2)

λE = flux density of latent heat into the air (W/m^2)

G = heat flux density into the water body (W/m^2)

To convert the above λE in W/m^2 into an equivalent evapotranspiration in units of mm/d, λE should be multiplied by a factor 0.0353. This factor equals the number of seconds in a day (86 400), divided by the value of λ (2.45 × 10^6 J/kg at 20°C). Density of water is assumed as 1000 kg/m^3. Penman (1948) combined energy balance approaches and turbulent transport of water vapor away from an evaporating surface.

$$E_0 = \frac{\left[{s}/{\gamma} (R_n - G) + E_a \right]}{{s}/{\gamma} + 1} \qquad \qquad ...(4.21)$$

where,

E_0 = evaporation from open water surface (mm/day)

R_n = net radiation in water equivalent (mm/day)

G = soil heat flux in water equivalent (mm/day)

$$R_n = R_A (1-\alpha)\left(0.32 + 0.46\frac{n}{N}\right)$$

$$-\sigma T_a^4 \left(0.55 - 0.92\sqrt{e_d}\right)\left(0.10 - 0.90\frac{n}{N}\right) \quad ...(4.22)$$

where,

α = Albedo

R_A = mean monthly radiation outside the atmosphere in water equivalent (mm/day)

T_a = mean air temperature ($^\circ$K)

s = slope of the saturation vapor pressure vs. temperature curve (mm Hg/$^\circ$C).

γ = psychometric constant (0.49 mm Hg/$^\circ$C)

n/N = ratio of actual to maximum possible sunshine hours

σ = Stefan-Boltzman constant = 5.56×10^{-8} w/m^2/K^4 = 2.06×10^{-9} mm/day

e_d = saturation vapor pressure of the atmosphere in mm Hg at dew point temperature or actual vapor pressure.

Again

$$e_d = \frac{MeanRH \times e_a}{100} = e_{sw} - 0.49\left(T_d - T_w\right) \quad ...(4.23)$$

where,

e_a = saturation vapor pressure at mean air temperature (mm Hg)

e_{sw} = saturation vapor pressure at wet bulb temperature (T_w) and dry bulb temperature(T_d) (mm Hg).

An aerodynamic component is explained by the following expression.

$$E_a = 0.35 (e_a - e_d) \times (1 + 0.0098u_2) \quad ...(4.24)$$

Here u_2 is wind speed in miles/day at 2 m height and u_2 can be determined as given below.

$$u_2 = \log 6.6 \times u_h / \log h \quad ...(4.25)$$

where,

u_h = wind speed at miles/day at height 'h' in feet

Example 4.7. Estimate the net radiation in water equivalent with the help of following given collected data in the field by using Penman combination method. This example is taken from Nivas et al. (2002).

Mean air temperature, $\qquad T_a = 16.1°C$

Mean relative humidity, $\qquad = 68.5\%$

Actual sunshine hrs , $\qquad n = 8.0$

Possible sunshine hrs, $\qquad N = 11.14$

Soil heat flux for water surface, $G = 0$

Wind speed at 10 feet height $\qquad = 17.6$ km/min = 10.92 miles/hr

$\qquad\qquad\qquad\qquad\qquad = 262.3$ miles/day

Mean solar radiation outside the atmosphere $(R_A) = 10.77$ mm of water

Saturation vapor pressure at mean air temperature, $e_a = 9.7$ mm Hg

Albedo, $\alpha = 0.25$

$s/\gamma = 1.769$ at 16.1°C

$\sigma T^4 = 14.05$ mm of water (can be taken from Appendix F)

Sol.: After carefully observing the data given, first determine the wind speed at 2 m height, since wind speed at 10 feet height is given. Using the following formula,

$$u_2 = \log 6.6 \times u_h / \log h$$
$$= \log 6.6 \times 262.3 / \log 10$$
$$= 219.47 \text{ miles/day}$$

Now, since saturation vapor pressure at mean air temperature is given as 9.7 mm Hg, saturation vapor pressure of the atmosphere in mm Hg at dew point temperature or actual vapor pressure, e_d can be calculated from the following formula,

$$e_d = \frac{Mean\ RH \times e_a}{100}$$
$$= \frac{68.5 \times 9.7}{100}$$
$$= 6.64 \text{ mm Hg}$$

By using of Eqn. 4.22, we can find out the net radiation in water equivalent (mm/day)

$$R_n = R_A(1-\alpha)\left(0.32 + 0.46\frac{n}{N}\right) - \sigma T_a^4\left(0.55 - 0.92\sqrt{e_d}\right)\left(0.10 + 0.90\frac{n}{N}\right)$$
$$-10.77(1\quad 0.25)\left(0.32 + 0.46 \times \frac{8}{11}\right) - 14.05\left(0.55 - 0.92\sqrt{6.64}\right)$$
$$\left(0.10 + 0.90 \times \frac{8}{11}\right)$$
$$= 5.26 - 4.39 \times 0.748$$
$$= 1.62 \text{ mm/day}$$

Finally, determine E_a with the given equation.

$$E_a = 0.35 \, (e_a - e_d) \times (1 + 0.0098 u_2)$$
$$= 0.35 \, (9.7 - 6.64) \times (1 + 0098 \times 219.47)$$
$$= 1.07 \times 3.15$$
$$= 3.37 \text{ mm/day}$$

By using Eqn. 4.21, evaporation from open water surface in mm/day can be determined.

$$E_0 = \frac{\left[1.769(1.62 - 0.0) + 3.37\right]}{1.769 + 1}$$

$$= (1.769 \times 1.62 + 3.37) \, / 2.769$$

$$= 3.85 \text{ mm/day}$$

Ans. 3.85 mm/day

4.1.5.2 Modified Penman Approach

Monteith (1963, 1965) introduced resistance terms into the Penman method and proposed the following equation to estimate latent heat of evaporation.

$$LE = \frac{\left[\left\{\frac{s}{\gamma}(R_n - G)\right\} + \left\{\rho_a C_p (e_s - e_a)/\gamma.r_a\right\}\right]}{\left(\frac{s}{\gamma} + 1 + \frac{r_c}{r_a}\right)} \quad ...(4.26)$$

where,

ρ_a = density of air (1.3 kg/m^3)

C_p = specific heat of air at constant pressure (1008 J/kg/$^\circ$C)

r_a = aerodynamic resistance (s/m)

r_c = canopy resistance (s/m)

    = $r_s + 15$

r_s = stomatal resistance (s/m)

$$r_s = \frac{\left[(r_{ad} \times r_{ab})/(r_{ad} + r_{ab})\right]}{LAI} \quad ...(4.27)$$

where,

r_{ad} = adaxial resistance

r_{ab} = abaxial resistance

LAI = leaf area index

e_a = actual vapor pressure (mm Hg)

e_s = saturation vapor pressure (mm Hg)

Aerodynamic resistance (r_a) is determined as follows.

$$r_a = \frac{\left[\ln\left\{(z-d)/z_0\right\}\right]^2}{uk^2} \qquad \qquad ...(4.28)$$

where,

z = height

d = zero plane displacement = $0.63z$

z_0 = roughness parameter = $0.13z$

u = wind speed at height, z

k = von Karman's constant (0.41)

Example 4.8. Determine the latent heat of evaporation by using modified Penman approach with the following data collected in the field.

Net radiation, R_n = 282.0 W/m^2

Soil heat flux, G = 10.1 W/m^2

s/γ = 1.860 at 17.5°C

Density of air, ρ_a = 1.3 kg/m^3

Specific heat of air at constant pressure,	C_p	= 1008 J/kg/°C
Psychometric constant,	γ	= 0.49 mm Hg/°C
Dry bulb temperature		= 17.0°C
Wet bulb temperature		= 9.8°C
Saturation vapor pressure,	e_s	= 5.0 mm Hg
Actual vapor pressure,	e_a	= 1.5 mm Hg
Wind speed at height (z = 1m),	u	= 0.40 m/s
Adaxial resistance,	r_{ad}	= 3.63 s/cm
Abaxial resistance,	r_{ab}	= 2.05 s/cm
Leaf area index,	LAI	= 4.0
Roughness parameter,	z_0	= 0.13z
von Karman's constant,	k	= 0.41

Sol.: Since z is given as 1 m, the zero plane displacement (d), which is equal to $0.63z$, will be 0.63 m.

Aerodynamic resistance (r_a) can be determined by using Eqn. 4.28

$$r_a = \frac{\left[\ln\left\{(1-0.63)/0.13\right\}\right]^2}{0.40(0.41)^2}$$

$$= 16.27 \text{ s/m}$$

The stomatal resistance (s/m) is determined by the following formula.

$$r_s = \frac{\left[\left(r_{ad} \times r_{ab}\right)/\left(r_{ad} + r_{ab}\right)\right]}{LAI}$$

$$= \frac{\left[\left(3.63 \times 2.05\right)/\left(3.63 + 2.05\right)\right]}{4.0}$$

$$= 0.33 \text{ s/cm} = 33 \text{ s/m}$$

Therefore, canopy resistance $(r_c) = r_s = 15$

$$r_c = 33 + 15 = 48.0 \text{ s/m}$$

Putting these values in Eqn. 4.26, latent heat of evaporation can be determined,

$$LE = \frac{\left[\left\{1.860\left(282 - 10.1\right)\right\} + \left\{1.3 \times 1008 \times \dfrac{5.0 - 1.5}{0.49 \times 16.27}\right\}\right]}{\left(1.86 + 1 + {48.0}/{16.27}\right)}$$

$$= (505.734 + 567.013) / (2.86 + 2.95)$$
$$= 184.6 \text{ W/m}^2$$
$$= 0.27 \text{ mm/hr}$$

Ans. 3.85 mm/day

4.1.5.3 Van Bavel method

Van Bavel (1966) modified the Penman method and introduced a new term 'B_V', and proposed the following formula.

$$LE = \frac{\left[s\left(R_n - G\right) + \gamma \times L \times B_V \times d_a\right]}{s + \gamma} \qquad \text{...(4.29)}$$

$$B_V = \left\{\frac{\varepsilon \rho_a k^2}{P}\right\} \times \left[\frac{u}{\left\{\ln\left\{{\left(z - d\right)}/{z_0}\right\}\right\}^2}\right] \qquad \text{...(4.30)}$$

where,

d_a = vapor pressure deficit (mm Hg) is equal to difference of e_s and e_a

P = air pressure, 0.75 mm Hg

u = wind speed (m/s)

z = crop height, m

d = zero plane displacement

L = latent heat, J/kg, 2.346

z_0 = roughness parameter

ε = 0.622 (it is a ratio of molecular weight of water to molecular weight of dry air)

4.1.5.4 Slatyer and Mcllroy Method

The Slatyer-Mcllroy equations are based on evapotranspiration from a wet surface with minimum advection. This happens when the air above a surface is saturated, due to vapor exchange with the wet surface. They used wet bulb depression instead of vapor pressure deficit and proposed the following modified equation.

$$ET_0 = \left[s \frac{(R_n - G)}{s + \gamma} \right] + \left(\frac{\rho_a C_p D_z}{r_a} \right) \qquad ...(4.31)$$

where,

D_z = wet bulb depression at height 'z'. It is a difference of dry and wet bulb temperature.

Example 4.9. Determine the potential evapotranspiration by using the Slatyer and Mcllroy method with the flowing data.

Net radiation	R_n	= 310.0 W/m^2
Soil heat flux	G	= 12.6 W/m^2
Slope of the saturation vapor pressure vs temperature curve,	s	= 0.91mm Hg/°C
Psychometric constant,	γ	= 0.49 mm Hg/°C
Density of air,	ρ_a	= 1.3 kg/m^3
Specific heat of air at constant pressure,	C_p	= 1008 J/kg/°C
Aerodynamic resistance,	r_a	= 14.30 s/m
Wet bulb depression	D_z	= 1.4°C

Sol.: Putting all the given values in Eqn. 4.31,

$$ET_0 = \left[s \frac{(R_n - G)}{s + \gamma} \right] + \left(\frac{\rho_a C_p D_z}{r_a} \right)$$

$$= \left[0.91 \times \frac{(310 - 12.6)}{0.91 + 0.49} \right] + \left(\frac{1.3 \times 1008 \times 1.4}{14.30} \right)$$

$$= 0.91 \ (212.42) + 128.29$$
$$= 193.30 + 128.29$$
$$= 321.59 \ \text{W/m}^2 = 0.48 \ \text{mm/hr}$$

Ans. 0.48 mm/hr

4.1.5.5 Priestley and Taylor Method

The Priestley-Taylor model (Priestley and Taylor, 1972) is a modification of Penman theoretical equation. An empirical approximation of the Penman combination equation is made by the Priestley-Taylor to eliminate the need for input data other than radiation. The adequacy of the assumptions made in the Priestley-Taylor equation has been validated by a review of 30 water balance studies in which it was commonly found that, in vegetated areas with no water deficit or very small deficits, approximately 95% of the annual evaporative demand was supplied by radiation (Stagnitti et al., 1989).

Priestley and Taylor (1972) found that the actual evapotranspiration from well watered vegetation was generally higher than the equilibrium potential rate and could be estimated by multiplying the ET_0 by a factor (α) equal to 1.26. They demonstrated that potential evapotranspiration is directly related to the equilibrium evaporation in the absence of advection and propose the following equation.

$$ET_0 = \frac{\alpha \times s \times (R_n - G)}{s + \gamma} \qquad \qquad \text{...(4.32)}$$

where, α is the empirical constant and is taken as 1.26 for temperature ranging between 15°C and 30°C. Although, the value of α may vary throughout the day, there is general agreement that a daily average value of 1.26 is applicable in humid climates (Shuttleworth and Calder, 1979). Morton (1983) observed that the value of 1.26, estimated by Priestley and Taylor, was developed using data from both moist vegetated and water surfaces.

Morton recommended α as 1.32 for estimation from vegetated areas as a result of the increase in surface roughness. Generally, the coefficient α for an expansive saturated surface is usually greater than 1.0. This means that true equilibrium potential evapotranspiration rarely occurs; there is always some component of advection energy that increases the actual evapotranspiration. Higher values of α, ranging up to 1.74, have been recommended for estimating potential evapotranspiration in more arid regions (ASCE, 1990).

Example 4.10. Estimate the potential evapotranspiration by using the Priestley and Taylor method. The following data were collected and given below.

Net radiation, R_n = 310.0 W/m^2

Soil heat flux, G = 12.6 W/m^2

Slope of the saturation vapor pressure vs temperature curve,
$$s = 0.91 \text{mm Hg/°C}$$
Psychometric constant, $\quad \gamma = 0.49$ mm Hg/°C

An empirical constant, $\quad \alpha = 1.26$

Sol.: All the data required by the Priestley and Taylor method of computing potential evapotranspiration are given, so it is easier to put all the values in Eqn. 4.32.

$$ET_0 = \frac{\alpha \times s \times (R_n - G)}{s + \gamma}$$

$$= \frac{1.26 \times 0.91 \times (310 - 12.6)}{0.91 + 0.49}$$

$$= 243.57 \text{ W/m}^2$$

$$= 0.36 \text{ mm/hr}$$

Ans. 0.36 mm/hr

4.1.6 Methods in Remote Sensing Technology

Canopy temperature can be determined with remote sensing and this is used in energy balance models which provide reliable estimates of evapotranspiration estimation. Thermal infrared radiation data recorded from radiometers or scanners provide an input to methods of estimation of ET_0. There are several methods such as Bartholic, Namken and Wiegand model, Brown and Rosenberg resistance model, stability corrected aerodynamic resistance model, energy balance–regression model, and Soer model, which are being used to estimate ET_0. We are not discussing these models in details but all the methods usually employed in remote sensing technology are basically based on energy balance equation which is expressed below.

$$R_n = H + E + G \qquad \qquad ...(4.33)$$

where,

$$R_n = \text{net radiation (W/m}^2\text{)}$$
$$H = \text{sensible heat flux (W/m}^2\text{)}$$
$$E = \text{latent heat flux (W/m}^2\text{)}$$
$$G = \text{soil heat flux (W/m}^2\text{)}$$

4.2 Crop Water Requirement in Greenhouse Environments

Inadequate irrigation tends to waste water, nutrients and energy, and may cause soil degradation by water-logging and salinisation. Under closed spaces such as greenhouses, the predominant role of crop transpiration in decreasing the heat load during warm periods is a supplementary reason to develop irrigation scheduling that allow the maximization of the transpirational fluxes. The main process involving the fate of water in the greenhouse is evapotranspiration, a process that is driven by a constant inflow of energy. As the ultraviolet stabilized plastic films, used in greenhouse construction, changes locally the radiation balance inside the greenhouse by entrapping the long wave radiation and creates a barrier to moisture losses, the crop water requirement is reduced due to less evapotranspiration.

Crop cultivation under greenhouse reduces evapotranspiration to about 70% of open field. It, thus, improves the water use efficiency relative to unprotected cropping (Stanghellini, 1993). It has also been reported that farming under greenhouse can save about 20–25% of water compared to the open drip irrigated farming systems. For representing the conditions inside the greenhouse, Stanghellini (1987) revised the Penman-Monteith model. Normally, wind speeds are typically less than 1.0 m/s. The Stanghellini model includes the crop canopy and aerodynamic resistance terms as well as more complex calculations of the solar radiation heat flux derived from the empirical characteristics of short wave and long wave radiation absorption in a multi-layer canopy.

A leaf area index is used to account for energy exchange from multiple layers of leaves on crops grown inside the greenhouse. The equation to estimate potential evapotranspiration (ET_0) takes the forms as given hereunder.

$$ET_0 = 2 \times LAI \cdot \frac{1}{\lambda} \frac{\Delta(R_n - G) + K_t \dfrac{VPD.\rho.C_p}{r_a}}{\Delta + \gamma\left(1 + \dfrac{r_c}{r_a}\right)} \qquad ...(4.34)$$

$$R_n = 0.07 R_{ns} - \frac{252.\rho.C_p(T - T_0)}{r_R} \qquad ...(4.35)$$

where,

$$\gamma = \frac{C_p.P}{\varepsilon.\lambda} \qquad ...(4.36)$$

$$r_R = \frac{\rho.C_p}{4.\sigma(T + 273.15)^3} \qquad ...(4.37)$$

LAI = leaf area index in m²/m²
K_t = time unit conversion (86400s/day for ET_0 in mm/day)
r_c = canopy resistance (s/m)
R_{ns} = net short wave radiation (MJ/m²/day)
T_0 = leaf temperature (°C)
r_a = aerodynamic resistance (s/m)
C_p = specific heat of air which equals to 0.001013 (MJ/kg/°C)

ρ = mean air density (kg/m^3)

ε = water to dry molecular weight ratio

λ = latent heat of vaporization (MJ/kg)

σ = Stefan-Boltzmann constant (MJ/m^2/K^{-4}/day)

r_R = radiative resistance (s/m)

P = atmospheric pressure (kPa)

VPD = vapor pressure deficit (kPa)

4.3 Computer Models to Estimate Evapotranspiration

4.3.1 ET$_0$ Calculator

Reference evapotranspiration (ET$_0$) is a climatic parameter and can be estimated from available weather data. It explains the evaporating power of the atmosphere at a specific site and does not consider the crop characteristics and soil factors. The ET$_0$ calculator software estimates reference evapotranspiration by using the FAO Penman-Monteith equation. This method closely approximates grass ET$_0$ and is a physically-based model which incorporates both physiological and aerodynamic parameters. The input data is classified in four groups such as air temperature, humidity, wind speed, and sunshine and radiation. You will need to create a file wherein you will be required to fill up the temperature and humidity data. If you do not have the wind speed data, you need to choose the options available such as light wind to strong wind.

The output data can be plotted and imported to some other crop growth simulation model such as AquaCrop model, also developed by FAO, directly as the input file. The ET$_0$ calculator was developed by Dr. Dirk Raes of Katholieke Universiteit Leuven, Belgium in cooperation with Land and Water Division of FAO. Students must download it from FAO website and practice it to observed how humidity and temperatures, wind speed are changing the plot of ET$_0$.

4.3.2 CROPWAT

CROPWAT 8.0 has been developed by Joss Swennenhuis for the Water Resources Development and Management Service of FAO. CROPWAT 8.0 is based on the DOS versions CROPWAT 5.7 of 1992 and CROPWAT 7.0 of 1999. CROPWAT 8.0 is a computer program for the estimation of crop water requirements from climatic and crop data. The program also allows the development of irrigation schedules for different management conditions and the calculation of water supply scheme for varying crop patterns. All calculation procedures as used in CROPWAT 8.0 are based on the FAO guidelines as laid down in the publication No. 56 of the Irrigation and Drainage Series of FAO "Crop Evapotranspiration—guidelines for computing crop water requirements". It can be easily downloaded from FAO website and takes directly the input of reference evapotranspiration calculated from ET$_0$ calculator.

4.4 Computation of Crop Water Requirement with Limited Wetting

In drip irrigation systems, only part of the soil surface is wetted and for widely spaced crops, crop canopy coverage is also limited. It is not appropriate to consider the soil evaporation from the entire soil surface under drip irrigation systems. Vermeiren and Jobling (1984) suggested using a correction factor (K_r) to take into the account of percentage of crop canopy coverage of cultivated land. The relationship can be expressed hereunder:

$$ET_{CROPCor} = K_r \times ET_{CROP} \qquad \qquad ...(4.38)$$

where, $ET_{CROPCor}$ is corrected crop water requirement, K_r is correction factor and ET_{CROP} is crop water requirement without considering limited area wetting. Keller and Bliesner (1990) developed the following formula to estimate crop evapotranspiration or crop water requirement for limited wetted areas.

$$ET_{CROPCr} = ET_{CROP}\left[0.1 \times \sqrt{P_d}\right] \qquad \qquad ...(4.39)$$

where, P_d is percentage of crop canopy coverage.

Example 4.11. A drip irrigated tomato crop is at development stage and the relative humidity is 70% with wind speed of 2 m/s. Crop canopy coverage is 60% and K_c value for development stage is 0.75. Take ET_0 as 5.2 mm/day and correction factor as 0.92. Compare the evapotranspiration estimation without area wetting, Vermeiren and Jobling (1984), and Keller and Bliesner (1990) method for limited area wetting.

Sol.: Let us estimate the crop evapotranspiration when limited area concept is not being applied.

$$ET_{CROP} = K_c \times ET_0$$
$$= 0.75 \times 5.2$$
$$= \textbf{3.9 mm/day}$$

When limited area concept is being applied, according to Vermeiren and Jobling (1984),

$$ET_{CROPCor} = K_r \times ET_{CROP}$$
$$= 0.92 \times 3.9$$
$$= \textbf{3.59 mm/day}$$

Finally, according to Keller and Bliesner (1990),

$$ET_{CROPCr} = ET_{CROP}\left[0.1 \times \sqrt{P_d}\right]$$
$$= 3.9 \,(0.1 \times \sqrt{60}\,)$$
$$= \textbf{3.02 mm/day}$$

5

IRRIGATION SCHEDULING

The increasing shortages of water and costs of irrigation are leading to develop methods of irrigation that maximize the water use efficiency. The advent of precision irrigation methods such as drip irrigation has played a major role in reducing the water required in agricultural and horticultural crops, but also has highlighted the need to develop new methods of irrigation scheduling. Irrigation scheduling aims to achieve an optimum supply of water to the crops for productivity with soil water content maintained at field capacity. Irrigation scheduling is the decision of when and how much water should be applied to the crops in a field. The purpose is to find out the exact amount of water to be applied to the field and the irrigation interval. It saves water and energy and thereby increases irrigation efficiency.

There are certain irrigation criteria which are used to determine when and how much water to be supplied. The most common criteria are soil moisture content and soil moisture tension. For example, suppose a farmer whose main objective is to maximize crop yield consider soil moisture content as the irrigation criterion. He decides that when soil water content drops below 80% of the total available soil moisture, irrigation should start. Soil moisture content to trigger irrigation depends on the farmer's goal, strategy, and water availability. The farmer will try to keep the soil moisture content above the critical level. Thus, irrigation is applied whenever the soil water content level reaches the critical level. This chapter deals with the question that when to irrigate the crop, how much of water will be needed that has been explained in the Chapter 4 on 'Estimation of Water Requirement of Crops'.

5.1 Advantages of Irrigation Scheduling
- Can supply water to the various fields to minimize crop water stress and maximize yields.
- It can make maximum use of soil moisture storage.
- Increasing water use efficiency means reducing fertilizer cost.

- Increases net returns by increasing crop yield and quality.
- Minimizes waterlogging and reduces drainage requirements.
- Controls root zone salinity problems.

5.2 Methods of Irrigation Scheduling

5.2.1 Water Balance Approach

The water balance approach to irrigation scheduling keeps the account of all water additions and subtractions from the soil root zone. As the crop grows and extracts water from the soil to satisfy its evapotranspiration requirement, the soil water storage is gradually depleted. The water requirement of the crop can be met by stored soil water, rainfall, and irrigation. Irrigation is required when crop water requirement exceeds the supply of water from soil water and rainfall. The water balance method of irrigation scheduling is a method which estimates the required amount and timing of irrigation for crops. This method needs the prior information on initial soil water content in the root zone, crop evapotranspiration, rainfall, and the available water capacity of the soil. The available water capacity or total available water of the soil is the amount of water available for plants to use in between the field capacity and permanent wilting point. The net irrigation requirement is the amount of water required to refill the soil water content of root zone to field capacity. This amount is the difference between field capacity and present soil water content and called the soil water deficit (D_C). On a daily basis, D_C can be determined by using the following equation:

$$D_C = D_P + ET_{CROP} - R - I_{RR} - U + S_{RUNOFF} + DP \qquad ...(5.1)$$

where,

$\quad\quad D_P$ = soil water deficit on the previous day,

$\quad ET_{CROP}$ = crop evapotranspiration rate for the current day,

$\quad\quad R$ = total rainfall for the current day,

$\quad\quad I_{RR}$ = net irrigation amount infiltrated into the soil for the current day,

$\quad\quad U$ = upflux of shallow groundwater into the root zone,

$\quad S_{RUNOFF}$ = surface runoff,

$\quad\quad DP$ = deep percolation or drainage.

The last three variables in Eqn. 5.1 (U, S_{RUNOFF}, DP) are difficult to estimate in the field. In many situations, the water table is significantly deeper than the root zone and U is assumed to be zero. Also, S_{RUNOFF} and DP can be accounted for in a simple way by setting D_c to zero whenever water additions (R and I_{RR}) to the root zone are greater than $D_P + ET_{CROP}$.

Using these assumptions, Eqn. 5.1 can be simplified to:

$$D_C = D_P + ET_{CROP} - R - I_{RR} \qquad \qquad ...(5.2)$$

If value of D_C is negative, then it can be set to zero. The amounts of water used in the equations are typically expressed in depths of water per unit area (e.g. inches of water per acre or mm of depth per hectare). Eqn. 5.2 is a simplified version of the soil water balance with several underlying assumptions. First, any water additions (R and I_{RR}) are assumed to readily infiltrate into the soil surface and the rates of R or I_{RR} are assumed to be less than the long-term steady state infiltration rate of the soil. Actually, some water is lost to surface runoff if precipitation or irrigation rates exceed the soil infiltration rate. Thus, Eqn. 5.2 will underestimate the soil water deficit or the net irrigation requirement if R or I_{RR} rates are higher than the soil infiltration rate.

Knowledge of effective precipitation ($R-S_{RUNOFF}-DP$) irrigation, and soil infiltration rates (e.g. mm per hour) are required to obtain more accurate estimates of D_C. Secondly, water added to the root zone from a shallow water table (U) is not considered. Groundwater contributions to soil water in the root zone must be subtracted from the right hand side of the equation in case of a shallow water table. Eqn. 5.2 will overestimate D_C, if any actual soil water additions from groundwater are neglected. In irrigation practice, only a percentage of available water content (AWC) is allowed to be depleted because plants start to experience water stress even before soil water is depleted down to permanent wilting point (PWP). Therefore, a management allowed depletion (MAD) of the AWC must be specified. The rooting depth and MAD for a crop will change with developmental stage. The MAD can be expressed in terms of depth of water (d_{MAD} in mm of water) using the following equation.

$$d_{MAD} = \frac{MAD}{100}.AWC.z \qquad \qquad ...(5.3)$$

where, *MAD* is management allowed depletion (%), *AWC* is available water capacity of the root zone (mm of water per mm of soil), and z is depth of root zone (mm). The value of d_{MAD} can be used as a guide for deciding when to irrigate. Typically, irrigation water should be applied when the soil water deficit (D_C) approaches d_{MAD}, or when $D_C \geq d_{MAD}$. To minimize water stress on the crop, D_C should be kept less than d_{MAD}. If the irrigation system has enough capacity, then the irrigator can wait until D_C approaches d_{MAD} before starting to irrigate. The net irrigation amount equals to D_C can be applied to bring the soil water deficit to zero. Otherwise, if the irrigation system has limited capacity (maximum irrigation amount is less than d_{MAD}), then the irrigator should not wait for D_C to approach d_{MAD}, but should irrigate more frequently to ensure that D_C does not exceed d_{MAD}.

Rational method is the simplest and oldest method for estimation of runoff which can be written as:

$$Q = CIA \qquad \qquad ...(5.4)$$

where,

Q = Peak runoff rate, m^3/s

C = Runoff coefficient

I = Rain intensity (mm/hr)

A = Area of the field contributing runoff (ha)

The upward movement of water through the soil from a water table to the root zone is commonly known as 'capillary rise' and for steady state flow, the rate of upward capillary rise, U (cm/s) is given by:

$$U = a.\frac{(d\psi/dz)-1}{\psi^n + b} \qquad \qquad ...(5.5)$$

where,

ψ = Water potential, cm

z = Height above water table, cm

a, b and n = constants. For saturated soil, when $\psi = 0$, hydraulic conductivity = a/b. For typical coarse and fine textured soils, n can be taken as 4 and 1.5, respectively (Gardner, 1958).

5.2.2 Soil Moisture Measurement

This is one of the most commonly used methods to determine timing of irrigation keeping in view the soil moisture depletion. As the crop grows, it uses the water available within the soil profile of its root zone. As the water is being extracted by the crops, the soil moisture reaches a threshold level at which irrigation is required. If water is not applied, the plant will continue further to use the available water in the soil and finally dies. When the soil profile is full of water, it means that soil profile is at 100% moisture content or at about 0.1 bars of tension. At this time we say that soil moisture is at field capacity (FC). Soil moisture tension is a measurement of how tightly the soil particles hold onto the water molecules in the soil: The tighter the hold, the higher the tension. At FC, with a tension of only 0.1 bars, the water is not being held tightly and it is easy for plants to extract water from the soil. As the water deplete due to the use by the plant, the tension in the soil increases.

Figure 5.1 shows the relationship between the water availability and soil water potential or tension for sand, clay and loam soils. The plants will continue to use the available water of the soil until the soil moisture level goes to the PWP. When the soil moisture goes beyond the PWP, plants will not be able to extract the water anymore and finally the plants die. Although there is still some moisture in the soil below the PWP, this water is held so tightly by the soil particles that it cannot be extracted by the plant roots. For most of the agronomical crops, PWP occurs at 15 bars. This means that the soil is holding on very tightly to the water in its pores. In order for plants to use this water, they must create a suction greater than 15 bars.

For most commercial crops, this is not possible. At 15 bars, most plants begin to die. The difference between field capacity and PWP is called the plant available water (PAW). The majority of irrigation research recommends irrigating row crops such as grain or cotton when the MAD approaches 50%. For vegetable crops, the MAD is usually set at 40% or less, because they are more sensitive to water stress. For drip irrigation system, normally MAD is 10–20%. These deficit amounts must insure that water stress will not cause any yield losses. Plant's available water must be monitored throughout the crop growing season, so that the appropriate timing of irrigation can be anticipated. The following approaches are used to determine soil moisture content.

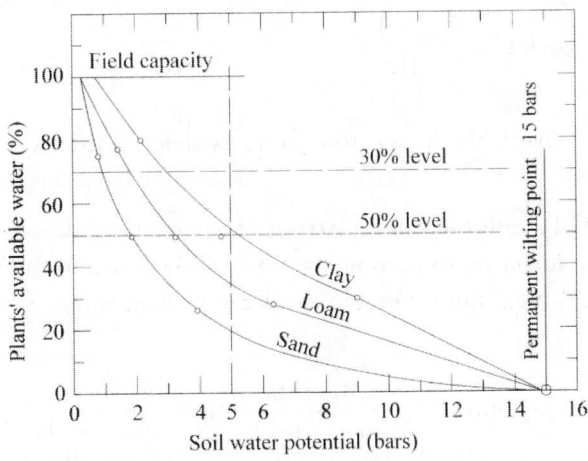

Fig. 5.1: Variations of available water with soil water potential for sand, clay and loam (National Engineering Handbook, 210-VI)

Field Capacity (FC)

The field capacity is interchangeably used with the terms water holding capacity and water retention capacity. This is the amount of soil moisture content held in soil after excess water has drained away and the rate of downward movement has materially decreased. The bulk water content retained in soil at −33 J/kg (or −0.33 bar) of hydraulic head or suction pressure is also called field capacity. It is represented by and in terms of equivalent depth, it can be expressed by:

$$FC = \frac{\theta_{FC}}{100} \times z \qquad \qquad ....(5.6)$$

where FC is the field capacity (mm), θ_{FC} is the field capacity of soil (%), and z is the crop root zone depth (mm).

Permanent wilting point (PWP): It is the point when there is no water available to the plant. PWP depends on the plant variety, but it is usually around 1500 kPa (15 bars). At this stage, the soil still contains some water, but it is difficult for the roots to extract it from the soil. It is also presented in percentage by volume (%).

Available water content: It is the amount of water which is actually available to the plants for their growth requirement. It is determined as FC minus the water that will remain in the soil at PWP. The available water content depends greatly on the soil texture and structure. The soil moisture at the available water capacity is expressed as given hereunder:

$$\theta_{AWC} = \theta_{FC} - \theta_{PWP}$$

where, θ_{AWC} is the maximum available moisture content (%).

The available water content (AWC) in (cm/cm) is determined as follows:

$$AWC = \frac{\theta_{FC} - \theta_{PWP}}{100} \qquad \qquad ...(5.7)$$

The total water available in the root zone is determined as $\frac{\theta_{FC} - \theta_{PWP}}{100} \times z$.

Depletion of available soil moisture: The percentage depletion of available soil water is the lowering of current state of soil moisture from FC with respect to the theoretical maximum possible available soil moisture. It is expressed as follows:

$$\text{Depletion, } \% = \frac{\theta_{FC} - \theta_0}{\theta_{FC} - \theta_{PWP}} \times 100 \qquad \qquad ...(5.8)$$

where, θ_0 is the current soil moisture content (%).

Management Allowable Depletion (MAD): In irrigation practice especially in microirrigation system, only a percentage of AWC is allowed to be depleted, because plants start to experience water stress even before soil water is depleted down to PWP. Therefore, MAD (%) of the AWC must be specified while scheduling irrigation. This is the maximum allowable percent depletion of the AWC. A 40% MAD refers to the fact that a maximum of 40 % of the AWC can be depleted before irrigation must take place. Therefore, MAD is defined as the percentage of total plant available water that is to be depleted from the active root zone before irrigation is applied (Table 5.1). The MAD can be expressed in terms of depth of water (mm) by the following expression:

$$MAD_{mm} = \frac{MAD}{100} \times \theta_{AWC} \times z \qquad \qquad ...(5.9)$$

Example 5.1. What is the depth of readily available water (RAW) for sandy clay loam if the effective root zone depth is 1.2 m, available water content as 0.30 and management allowable deficit is 0.2?

Sol.:

$$\text{Readily available water} = \theta_{AWC} \times z \times MAD$$
$$= 0.30 \times 1.2 \times 0.20$$
$$= 0.072 \text{ m} = 7.2 \text{ cm}$$

Ans. 7.2 cm

Example 5.2. If gravimetric water content is 30% and bulk density, ρ_b, is 1.45 g/cm³, then what is the volumetric water content and the porosity? Average particle density of soil, ρ_d, is 2.66 g/cm³.

Sol.:

$$\text{Volumetric water content} = \text{Gravimetric water content} \times \text{Bulk density}$$
$$= 0.30 \times 1.45 = 0.43 = 43\%$$

$$\text{Porosity, } \phi = 100 - \left(\frac{\rho_b}{\rho_d}\right) \times 100$$

$$= 100 - \left(\frac{1.30}{2.66}\right) \times 100 = 49\%$$

Ans. 43%, 49%

Example 5.3. What is the percent depletion, if measured water content is 15%, field capacity is 30% and permanent wilting point is 10%? If the MAD is 40%, at what water content must the next irrigation take place? If the root zone depth is 1.6 m, then what will be readily available water?

Sol.:

$$\text{Depletion, } \% = \frac{\theta_{FC} - \theta_0}{\theta_{FC} - \theta_{PWP}} \times 100 = \frac{30 - 15}{30 - 10} \times 100 = 75\%$$

where, θ_0 is the current soil moisture content (%).

The next irrigation must take place before the soil reaches a water content of 6%, i.e. 40% of AWC (30%–15%).

$$\text{Readily available water (RAW)} = AWC \times z \times MAD$$
$$= 0.15 \times 1.6 \times 0.40 = 0.096 \text{ m} = 9.6 \text{ cm}$$

Ans. 75%, 6%, 9.6 cm

Table 5.1: Maximum/management allowable depletion (MAD) and rooting depth for crops (FAO, 1989)

Crop	MAD (%)	Maximum root depth (cm)	Total growing period of crop (days)
Beans (dry)	40	90	90–120
Beans (green)	50	90	60–90
Corn (grain) or maize	50	60–90	90–110
Corn (sweet)	65	120	90
Onion (dry)	50	60	120
Onion (green)	50	60	90
Pasture/turf	60	60	65
Peas	40	60	100
Potatoes	30	60	90–120
Safflower	65	180	—
Sorghum (jowar)	65	60–90	135
Soybean	65	90	90–140
Sunflowers	65	90–120	—
Wheat	50	90	120
Cotton	50	120–150	195
Paddy or rice	70	30–60	120
Groundnut	60	60–75	120
Gram	50	120–150	110
Mustard	45	120–150	100
Sugarcane	60	120	365

5.2.2.1 *Neutron Probe*

This is a device used to measure the amount of moisture present in the soil. A neutron probe contains americium-241 and beryllium-9. The alpha particles are emitted by the decay of the americium and collide with the light beryllium nuclei, thereby, producing fast neutrons with a spectrum of energies ranging up to 11 mega electron volt. When these fast neutrons hit the hydrogen atom present in the soil, they lose their energy and slow down. Hydrogen atoms have the property of slowing down the speed of fast moving neutrons. A detector within the probe measures the rate of fast neutrons leaving and slow neutrons returning. The detection of slow neutrons returning to the probe allows an estimate of the amount of hydrogen present. Since water contains two atoms of hydrogen per molecule, this therefore gives a measure of soil moisture.

The major components of neutron probe are a probe, probe carrier, rate scaler and connecting cables (Fig. 5.2). When the probe is lowered down through the access tube at pre-determined depth, fast neutrons emit in the soil which collides with the hydrogen atoms present in the soil moisture and it gets scattered. The density of scatter cloud of fast moving neutrons is a function of soil moisture content. This cloud is sensed by the sensor in the probe and electrical pulse is transmitted to the rate scaler connected through the cable. The rate scaler displays the pulses in terms of counts per second.

These counts per second are converted into volumetric soil water content. As we know that some of the elements available in the soil have the scattering properties for fast moving neutrons, their impact cannot be ignored. Therefore, before using neutron probe, it is essential to calibrate the neutron probe. It accurately determines the soil moisture and is not affected by the temperature, soil types and pH. As neutron probe is very costly instrument and needs to have a licensed operator, this is usually bought by large organizations or research institutions. Soil moisture content using neutron probe is calculated as:

$$M_V = \frac{R_S}{R_{STD}} \times b - j \qquad \qquad ...(5.10)$$

where,

M_V = Volumetric moisture content (cm)

R_S = Observed counts/minute in soil

R_{STD} = Standard counts/minute in the field

b and j are calibration factors which are calibrated by plotting a graph between count ratio and moisture content in soil determined by gravimetric method in the field.

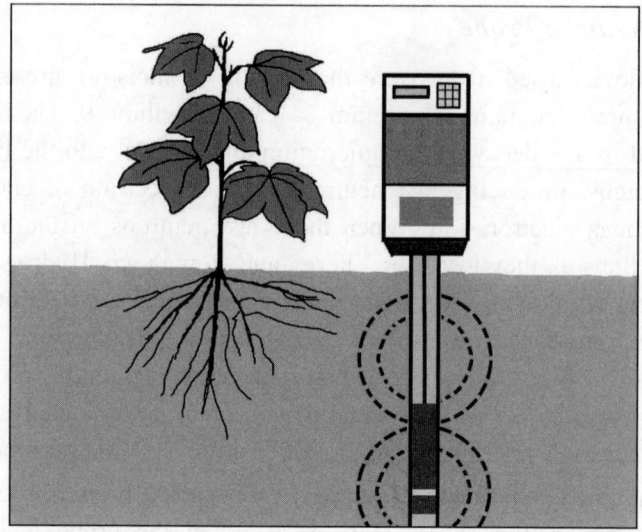

Fig. 5.2: A neutron probe inserted in the ground (Martin, 2009)

Example 5.4. Determine the soil moisture content using the neutron probe data as given below:

$$R_s = 4500 \text{ counts/min}$$
$$R_{std} = 6000 \text{ counts/min}$$
$$b = 0.35$$
$$j = 0.0$$

Sol.: Applying the Eqn. 5.6, we get

$$M_V = \frac{4500}{6000} \times 0.35 - 0.0$$

$$= 0.26 \text{ cm}$$

Ans. 0.26 cm

5.2.2.2 Electrical Resistance

Another method that has been used for several years to determine soil moisture content is electrical resistance. Devices such as gypsum blocks and Watermark sensors use electrical resistance to measure soil moisture. The principle behind these devices is that moisture content can be determined by the resistance between two electrodes embedded in the soil. The electrical resistances increase with decrease in soil moisture content. To measure soil moisture, the porous blocks containing electrodes are buried in the ground at the desired depth, with wire leads to the soil surface. Wheatstone electrical bridge is used to measure the high values of electrical resistance.

Fig. 5.3: Three resistance blocks anchored by a stake in the field (Martin, 2009)

The instrument can work for a wide range of soil moisture. A meter is connected to the wire leads and a reading is taken (Fig. 5.3). As the soil moisture changes, the water content of the porous block also increases or decreases and this change influences the electrical conductivity. Higher water content leads to higher conductivity or lower electrical resistance. The relationship between electrical resistance and soil moisture is quantified by a calibration process. This method is not suitable for soils with freezing temperatures and high salinity.

5.2.2.3 *Soil Tension*

When the soil dries out, the soil particles retain the water with greater force. Tensiometers measure how tightly the soil water is being held. Most tensiometers have a porous or ceramic tip connected to a water column. The tensiometers are installed to the desired depth (Fig. 5.4). As the soil dries, it begins to pull the water out of the water column through the ceramic cup, causing suction on the water column. This force is then measured with a suction gauge connected through a tube to porous cup. When the tensiometers are installed in the soil, the water in the porous cup reaches equilibrium with the moisture content in the surrounding soil. As the soil moisture decreases, the soil dries and the water begins to flow out of the cup.

The vacuum created in the cup by this suction of water is recorded on the pressure or suction gauge. In short, the changes in the soil tension reflect the changes in the soil moisture surrounding the cup. Some newer models have replaced the suction gauge with an electronic transducer. These electronic devices are usually more sensitive than the gauges. Tensiometers work well in soils with high soil-water content, but tend to lose good soil contact when the soil becomes too dry. The instruments also suffer from time-lag in response to changes in soil moisture.

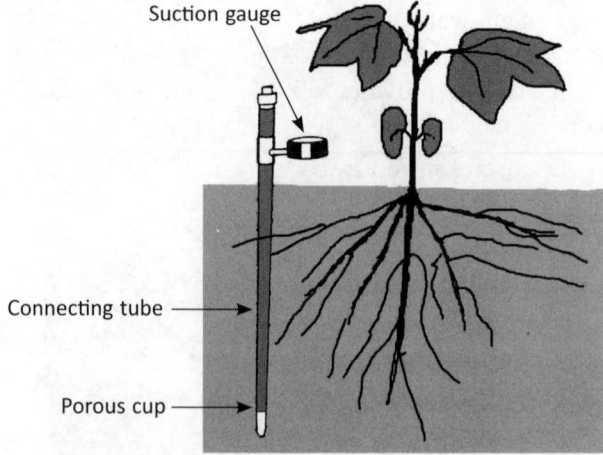

Suction gauge

Connecting tube →

Porous cup →

Fig. 5.4: A tensiometer in the field (Martin, 2009)

5.2.2.4 Time-Domain Reflectometry (TDR)

The TDR instruments work on the principle that the presence of water in the soil affects the speed of an electromagnetic wave. It slows down the speed of electromagnetic wave. The TDR sends an electromagnetic wave through a guide placed into the ground at the desired depth. It then measures the time taken by the wave to travel down the guide and reflect back to the guide. The time is recorded and converted to soil moisture reading. The wetter the soil, the longer it takes for the electromagnetic wave to travel down the guide and reflect back. The probes are connected to the instrument through a network of cables and multiplexers. The components of the TDR instrument include the voltage step and a fast oscilloscope which captures the reflected waveform. The oscilloscope can capture waveforms that represent all, or any part of, the waveguide. The relative height of the waveform represents a voltage which is proportional to the impedance of the waveguide. Most of the TDR instruments display the horizontal axis which actually measures in units of time. The TDR instrument converts the time measurement to length units by using the relative propagation velocity factor setting, v_p, which is a fraction of the speed of light in a vacuum. For a given cable, the correct value of v_p is inversely proportional to the permittivity, ε, of the dielectric (insulating plastic) between the inner and outer conductors of the cable and can be expressed below.

$$v_p = \frac{v}{v_c} = (\varepsilon\mu)^{-0.5} \qquad \qquad \text{... (5.11)}$$

where, v is the propagation velocity of the pulse along the cable, v_c is the speed of light in a vacuum, and μ is the magnetic permeability of the dielectric material. For a TDR probe in a soil, the dielectric between the probe rods is a complex mixture of air, water and soil particles that exhibits a variable apparent permittivity. For most soils, excluding those which are very high in organic matter, the TDR method provides water content in the range from zero to $0.5 \text{ m}^3/\text{m}^3$ with accuracy of 0.01 to 0.02 m^3/m^3 without calibration.

It is also very useful in root water uptake studies where information from discrete parts of the root zone is desired. Because TDR accurately integrates soil water content changes occurring along the length of the probe rods, TDR probes may be inserted vertically into soils to accurately assess mean water content over the length of the rods, even in soils exhibiting sharp water content changes with depth.

5.2.2.5 Frequency-Domain Reflectometers (FDR)

The neutron probe method as described earlier has the disadvantages of radioactive hazard, lack of automated data collection methodology, and involvement of high cost. The time domain reflectometry (TDR) method involves measuring the propagation of an electromagnetic pulse along the wave guides. By measuring the travel time and the velocity, the apparent dielectric constant of the soil can be estimated. Usually, the TDR method is not soil-specific and therefore no soil calibration is required and TDR measurements may be affected by soil salinity, soil temperature, clay type and clay content. The TDR technique may overestimate soil-water content in saline soils because the apparent dielectric constant also depends on the electrical conductivity of the soil. The frequency-domain reflectometer (FDR) method makes use of radio frequencies and the electrical capacitance of a capacitor, formed by using the soil and embedded rods as a dielectric, for determining the dielectric constant and thus, the soil water content. The signal reflected by soil combines with the generated signal to form a standing wave with amplitude that is a measure of the soil-water content.

In the case of capacitance-type sensors, such as that used by Grooves and Rose (2004), the charge time of a capacitor is used to determine the soil-water content. Profile-probe versions using FDR and capacitance methods are now commercially available (Mwale et al., 2005). The FDRs use an AC oscillator to form a tuned circuit with the soil. After inserting probes that are either parallel spikes or metal rings into the soil, a tuned circuit frequency is established. This frequency changes depending on the soil moisture content. Most models use an access tube installed in the ground. They read only a small volume of soil surrounding the probes. FDR is also sensitive to air gaps between the access tube and the soil. Many of these newer instruments require professional installation to operate properly.

5.2.2.6 Infrared/canopy Temperature

Plant indicators are also useful in determining the timing of irrigation. Observing few plants characteristic can give you a good idea of the status of the field's moisture content. An infrared (IR) thermometer measures the thermal temperature of the plant leaves or a crop canopy. As we know that plants transpire through openings called stomata, once the plants go into water stress, they begin to close their stomata and cease to transpire, causing the plant to heat up and the canopy temperature to rise. Infrared readings can detect this increase in plant temperature.

In this method, baseline temperatures need to be taken prior to measurements. The baseline temperature should be taken in a well-watered field, free of water stress. On days when the air temperature is very high, some plants will stop transpiring for a brief period. If infrared readings are being taken at that time, they may read that there is a water stress when, in fact, it is just a normal shutdown period. Compare field readings with your well-watered readings to make your decision. IR also requires taking temperature readings on clear days at solar noon. This normally occurs between noon and 2:00 p.m. This is to assure that the measurement you are taking is at maximum solar intensity. During the monsoon season, this may be difficult to achieve due to cloud cover. Early in the season, IR readings will often measure soil temperature when canopy cover is sparse. These readings usually result in higher temperature readings since the soil tends to heat up quickly. Figure 5.5 is a diagram of a hand-held IR gun.

Fig. 5.5: Hand-held infrared sensor (Martin, 2009)

5.2.3 Wetting Front Detector

At present numerous methods and devices are available for scheduling of irrigation. These include either physical measurement of the soil water content by means of, e.g. tensiometers and neutron probes, or simulation models using data from automatic weather stations and crop growth parameters. However, in practice it is found that irrigation scheduling is not widely applied mainly because methods are not always user-friendly and farmers have insufficient knowledge of scheduling tools. Dr. Richard John Stirzaker, a Scientist at CSIRO, Division of Land and Water, Australia developed a wetting front detector which also received the prestigious WatSave Technology Award in 2003 given by the International Commission on Irrigation and Drainage (Fig. 5.6).

The Wetting Front Detector is made of a funnel, a filter and a float, buried in the root zone. The detector works on the principle of flow line convergence. Irrigation water or rain moving downwards through the soil is concentrated when the water molecules enter the wide end of the funnel. The soil in the funnel becomes wetter as the funnel narrows and the funnel shape has been designed so that the soil at its base reaches saturation. Once saturation has occurred free water flows through a filter into a small reservoir and activates a float (Fig. 5.7). It gives a visual signal when the wetting front reaches the required soil depth.

The wetting front detector can be used to schedule irrigation because the time it takes for water to reach a certain depth depends on the initial water content of the particular soil. If the soil is dry before irrigation the wetting front moves slowly because the water must fill the soil pores on its way down. Therefore, a lot of water is needed before the detector will respond. Free water produced at the base of the funnel by convergence activates the float in the detector. Water is withdrawn from the funnel by capillary action after the wetting front dissipates.

Fig. 5.6: Dr Richard John Stirzaker holding the wetting front detector

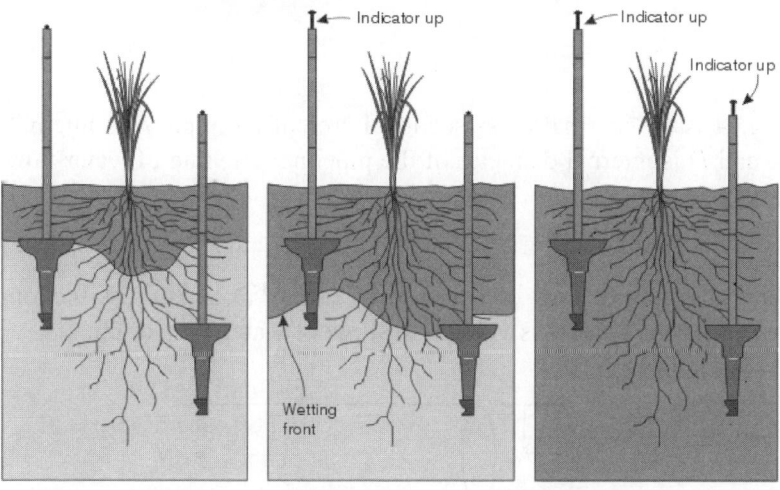

Fig. 5.7: Wetting front detector installed at different depths

6

HYDRAULICS OF WATER FLOW IN PIPES

The designing of any irrigation systems involves the determination of pipe sizes and uniformity of application of water in the field. The channel width of conventional irrigation systems is ensured to carry the necessary water required for irrigation. It includes all the losses, e.g. percolation losses, evaporation losses and leaching of water beyond root zone. Drip irrigation system also involves determination of pipe sizes to carry the desired amount of water. The amount of water to be carried depends upon the type of crop and climatic conditions of the region. Drip irrigation system is a network of pipes of various sizes; therefore, water flowing through pipe and its basic hydraulic is necessary to understand first.

6.1 Flow in a Pipe

The cross-sectional area of pipe is determined by

$$A = \pi R^2 = \pi \frac{D^2}{4} \qquad \qquad ...(6.1)$$

where, A is the internal cross-sectional area of the pipe, R is internal radius of pipe and D is internal diameter of the pipeline. The rate of water flow in a pipe can be given as

$$Q = A \times V \qquad \qquad ...(6.2)$$

where, Q is rate of flow of water (m³/s or cm³/s), A is cross-sectional area of pipe (m² or cm²) and V is velocity of flow of water (m/s or cm/s).

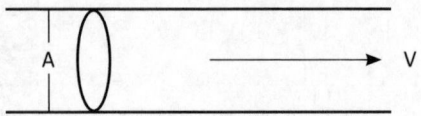

The velocity of flow of water in a pipe is given as,

$$V = L\!\!\Big/\!\!t$$

...(6.3)

where, L is the length of the pipe in which water flows and t is time.

Example 6.1. A drip irrigation system was installed in a banana orchard. A mainline of 63 mm was installed from A to C and sub-mainline of diameter 50 mm was taken out from point B. If the velocity of flowing water between A to B is 1.2 m/s and from B to C is 0.7 m/s, determine rate and velocity of flowing water coming out of B.

Sol.:

The cross-sectional area of mainline $A1 = \pi \dfrac{D^2}{4} = \pi \times \dfrac{6.3^2}{4} = 31.15$ cm^2

The cross-sectional area of sub-mainline $A2 = \pi \dfrac{D^2}{4} = \pi \times \dfrac{5^2}{4} = 19.62$ cm^2

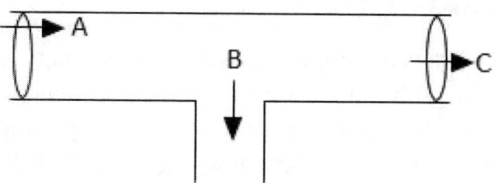

Since the flowing velocity of water is different for A to B and B to C, we will determine the discharge flowing through it.

For A to B:

$$Q_1 = A_1 \times V_1 = 31.15 \times 120 = 3738 \text{ cm}^3/\text{s}$$

For B to C:

$$Q_2 = A_2 \times V_2 = 19.62 \times 70 = 1373.40 \text{ cm}^3/\text{s}$$

The difference between the discharges Q_1 and Q_2 will pass through the sub-mainline, so suppose discharge passing out of sub-mainline is Q_3,

$$Q_3 = Q_1 - Q_2 = 3738 - 1373.40 = \mathbf{2364.60 \text{ cm}^3/\text{s}}$$

Cross-sectional area of sub-main line is already known, therefore, velocity of flowing water through sub-mainline is,

$$V = Q_3 \!\!\Big/\!\! A_2 = 2364.60 \!\!\Big/\!\! 19.60 = 120.64 \text{ cm/s} = \mathbf{1.20 \text{ m/s}}$$

Ans. 2364.60 cm^3/s, 1.20 m/s

6.2 Water Pressure

Water pressure describes the flow strength of water flowing through a pipe or other type of channel. It depends on water flow and more water flow will create more pressure. Water pressure can be expressed in meters rather than atmosphere or psi. The relationship can be shown as follows.

$$H = P/\gamma \qquad\qquad ...(6.4)$$

where, H is pressure head (m), P is in atmosphere and γ is specific weight (kg/cm^3). This can be clearer with this example.

Example 6.2. What will be the pressure of 2 atmospheric pressure?

Sol.: It is give that $p = 2$ atmosphere or in short atm. γ for water is 0.001 kg/cm^3.

Therefore, $H = P/\gamma = 2/0.001 = 2000$ cm = **20 m**

So, 2 atmosphere is 20 m of pressure head.

6.3 Estimation of Total Head

Bernoulli's principle states that for an inviscid flow, an increase in the fluid speed occurs simultaneously with a decrease in pressure or a decrease in the potential energy of fluid. Bernoulli's principle is based on the principle of conservation of energy. This means that, in a steady flow, the sum of all forms of mechanical energy in a flowing fluid is the same at all points. This requires that the sum of kinetic energy and potential energy remain constant. Total head or energy head H can be shown as follows.

$$H = z + \frac{p}{\rho g} + \frac{v^2}{2g} = h + \frac{v^2}{2g} \qquad\qquad ...(6.5)$$

Bernoulli's equation is based on certain assumptions which are stated below.

- Flow is steady.
- Density is constant, i.e. the fluid is incompressible.
- Friction losses are negligible.

Bernoulli's equation ceases to be valid before zero pressure is reached. The above equation uses a linear relationship between flow speed squared and pressure. The total head of water along the pipeline depends on the difference of elevation, pressure and velocity of water and it remains constant throughout the water body. For any two points along the flow, the equation takes the following form.

$$\frac{p_1}{pg} + \frac{v_1^2}{2g} + z_1 = \frac{p_2}{pg} + \frac{v_2^2}{2g} + z_2 \qquad\qquad ...(6.6)$$

where,

   z = the relative elevation of water (m)

   $\dfrac{p}{\rho g}$ = the pressure head of water (m)

   $\dfrac{v^2}{2g}$ = velocity head of flowing water (m)

Example 6.3. What is the hydraulic head and total energy of water in a pipe that is 5 m above the datum with pressure 350 kPa and water velocity 1.5 m/sec?

Sol.: Velocity head $= \dfrac{v^2}{2g} = \dfrac{1.5^2}{2 \times 9.81} = 0.114$ m

   Pressure head $= 350$ kPa $= 350,000$ Pa $= \dfrac{350000}{9.81 \times 1000} = 35.7$ m

   Hydraulic head $= 35.7 + 5 = 42.7$ m

   Total energy $= 42.7$ m $+ 0.114$ m $= 42.8$ m

Ans. 42.7 m, 42.8 m

6.4 Head Loss due to Friction in Plain Pipes

6.4.1 Hazen-William Formula

This is the most commonly used formula for estimating frictional head loss in a steady pipe flow of various pipe materials. The Hazen-Williams formula gives accurate head loss due to friction for fluids with kinematic viscosity of approximately 1.1 cSt [1 cSt (centiStokes) = 10^{-6} m^2/s]. It can be applied for pipes of 50 mm or larger and velocities less than 3 m/s. As the roughness coefficient used in this formula does not involve Reynolds number, and hence this gives the direct solutions. Due to its simplicity, Hazen-Williams formula has been extended to include plastic drip irrigation pipes even though they are smaller in diameter with lower flow rates than those normally encountered in other irrigation situations. The equation is acceptable for cold water at 15.6 °C. Formula is given below.

$$h_f = 1.212 \times 10^{12} . \frac{(Q/C)^{1.852}}{D^{4.87}} . \frac{L}{100} \qquad\qquad ...(6.7)$$

where,

   h_f = frictional head loss in the pipeline (m)

Q = discharge in the pipeline (lps)

C = friction coefficient to consider pipe material

L = length of the pipeline (m)

D = inside diameter of the pipeline (mm)

Manufacturers of pipes for drip irrigation recommended a C value of 150 for the plastic pipe and tubing. However, Watters and Keller (1978) showed that $C = 150$ underestimates the pipe friction losses for the flow normally encountered in drip irrigation systems. Howell and Hiler (1974) suggested that $C = 130$ be used for plastic pipes having diameter less than 16 mm. Howell et al. (1982) suggested that the best C values for drip irrigation systems are $C = 130$ for 14 –15 mm pipe diameter, $C = 140$ for 19–20 mm, and $C = 150$ for 25–27 mm.

The outside diameter (OD) of the pipe is the distance between the outside walls of the pipe, measured perpendicularly to the axis of the pipe. The inside diameter (ID) of the pipe is the distance between the internal walls of the pipe, measured perpendicularly to the pipe's axis as shown in Fig. 6.1. Wall thickness (WT) is the thickness of pipe. Pipe's inside diameter (ID) = OD – (2 × WT), and pipe's wall thickness (WT) can be given as (OD – ID) / 2.

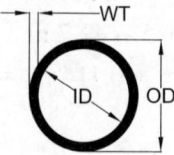

Fig. 6.1: Pipe's inside diameter

6.4.2 Darcy-Weisbach Formula

This is a dimensionally homogeneous formula, which is used for estimating head or pressure loss due to friction along a given length of pipe to the average velocity of the fluid. The Darcy-Weisbach equation contains a dimensionless friction factor, known as the Darcy friction factor. This is also called the Darcy-Weisbach friction factor or Moody friction factor. The Moody diagram shows the relationship between the relative roughness of the pipe and the Darcy-Weisbach friction factor (f) for different values of Reynolds number (R_n). The Darcy-Weisbach formula which relates the friction factor (f) to the head loss in pipes due to friction is given as below.

$$h_f = f \cdot \frac{L}{D} \cdot \frac{V^2}{2g} \qquad \qquad ...(6.8)$$

where,

h_f = head loss due to friction (m)

L = length of the pipe (m)

D = internal diameter of the pipe (mm)

V = average velocity of the fluid flow (m/s)

g = acceleration due to gravity (9.81 m/s^2)

f = a dimensionless coefficient called the Darcy friction factor

Blasius equation for smooth pipes has been recognized as an accurate predictor of friction factor for plain small diameter plastic pipe. However, in drip irrigation laterals with inserted emitters, the flow regime becomes semi-smooth due to the presence of emitter protrusions (Amin, 1990). Depending upon the flow conditions, the Darcy-Weisbach friction factor (f) can be calculated as fallows;

(a) For laminar flow (R_n< 2000), where pipe roughness is not a factor.

$$f = {}^{64}\!\big/\!{}_{R_n} \qquad\qquad ...(6.9)$$

(b) For turbulent flow ($4000 < R_n < 10^5$) in hydraulically smooth pipe such as glass, copper and plastic tubing, the Blasius equation can be used.

$$f = {}^{0.3164}\!\big/\!{}_{R_n{}^{0.25}} \qquad\qquad ...(6.10)$$

where, R_n is the Reynolds number. Howell and Barinas (1980) expressed Hazen-Williams formula in a form of Darcy-Weisbach equation to identify friction factor (f) as a function of roughness coefficient (C). A reference temperature of 15.6°C was used and the following expression was obtained:

$$f = \frac{1862}{C^{1.852}D^{0.18}R_n{}^{0.148}} \qquad\qquad ...(6.11)$$

where, D is in mm.

Reynolds number (R_e) is a dimensionless quantity and is used to determine whether a flow is laminar, transient or turbulent. It will tell you the mixing behavior of the fluid. Higher momentum (larger pipe diameter and higher velocity) and Reynolds number tends to propagate turbulent eddies while higher viscosity (lower Reynolds number) dampens out eddies and leads to laminar flow. Reynolds number is a ratio of inertial forces to viscous forces. For flow to be laminar, viscous flow will be dominant. It is expressed by

$$R_e = \frac{\rho VL}{\mu} \qquad\qquad ....(6.12)$$

where

ρ = density of the fluid (SI unit Kg/m^3)

V = velocity of fluid with respect to the object (m/s)

L = characteristic linear dimension (m)

μ = the dynamic viscosity of the fluid (Pa-s or N-s/m^2 or Kg/m-s)

Note: 1 cP = 0.01 poise = 0.01 gram per cm second = 0.001 Pascal second = 1 milli Pascal second = 0.001 N.s/m^2

The general form of the Reynolds number is also given by

$$R_e = \frac{\rho D v}{\mu}$$

where,

D = diameter of the flow channel (m)

v = the average velocity of the fluid in the channel (m/s)

ρ = fluid density (kg/m^3)

μ = viscosity of the fluid (kg/m.s)

6.4.3   Watters and Keller Formula

In an experiment conducted by Watters and Keller (1978), it was demonstrated that the Darcy-Weisbach formula can be used for estimating frictional head losses in smooth plastic pipes and tubes. They proposed a simplified form of Darcy-Weisbach formula which incorporates a friction factor estimated from the Blasius equation for smooth pipes with a water temperature of 20°C and kinematic viscosity of 1.0×10^{-6} m^2/sec.

$$h_f = 0.465 \frac{L}{D^{4.75}} Q^{1.75} \qquad\qquad ...(6.13)$$

where,

h_f = pipe friction loss (m)

L = pipe length (m)

D = inside pipe diameter (mm)

Q = pipe flow rate, liters/hr

6.5   Head Loss due to Friction in Multi-outlet Pipes

The drip irrigation lateral lines and sub-mains are the pipes with multiple outlets discharging laterally. The flow conditions in such pipes are generally considered as steady, spatially varied with decreasing discharge. Flow of water through the length of a single outlet (plain) pipe of a given diameter causes more friction loss than does flow through a pipe with a number of outlets. This introduces complications in estimation of frictional losses in multi-outlet pipes. The estimation of head loss caused by friction in pipelines with multiple outlets requires a stepwise analysis starting from most downstream outlet, working upstream and computing head loss caused by friction in each segment.

Christiansen (1942) developed a friction factor known as '*F*' factor, to avoid the cumbersome stepwise analysis. Computing the head loss in a pipe considering the entire discharge to flow through the entire length and multiplying by factor *F* allows the head loss through a single diameter pipeline, with multiple outlets to be estimated. The derivation of factor *F* was based on the following assumptions:

- No outflow at the downstream end of the pipeline.
- All outlets are equally spaced and have equal discharges.
- The distance between the pipe inlet and the first outlet is equal to one full outlet spacing.
- Hydraulic characteristics remain constant along the length of the pipeline.

Factor *F* is a dimensionless factor and a function of the friction formula used, and the number of outlets along the pipeline. The head loss in a pipeline without outlets can be calculated using any of the well-known friction formulas, such as Darcy-Weisbach, Hazen-Williams, and others. This is then multiplied by the factor *F* to calculate the head loss caused by friction in a pipeline with multiple outlets. Christiansen's equation for computing reduction coefficient '*F*' for multiple outlet pipelines, where the first outlet is at full spacing from the main line, can be written as,

$$F = \frac{1}{b+1} + \frac{1}{2N} + \frac{(b-1)^{0.5}}{6N^2} \qquad ...(6.14)$$

where,

b = velocity or flow exponent in the head loss equation used.

N = number of outlets in the pipeline.

In many field layouts, the first outlet on a pipeline cannot be located at a full spacing from the inlet of the pipe. Jensen and Fratini (1957) developed an adjusted factor *F* which permitted the calculation of the head loss caused by friction in pipelines with multiple outlets, with the first outlet at one-half of outlet spacing from the pipeline inlet. Scaloppi (1988) derived an expression for the adjusted factor F_a. This expression allows the adjusted factor to be calculated for a pipeline with multiple outlets and the first outlet at any fraction of spacing (α) from the pipeline inlet. If the first outlet is one-half of outlet spacing from the pipeline inlet, the adjusted factor by Scaloppi (1988) is identical to that by Jensen and Fratini (1957). Scaloppi (1988) derived adjusted factor F (α) can be expressed as:

$$F(\alpha) = \frac{NF - (1-\alpha)}{N - (1-\alpha)} \qquad ...(6.15)$$

and in the special case where the first outlet is at half of outlet spacing from the mainline ($\alpha = 1/2$)

$$F(\tfrac{1}{2}) = \frac{2N}{2N-1}\left(\left(\frac{1}{b+1}\right) + \frac{(b-1)^{0.5}}{6N^2}\right) \qquad \text{...(6.16)}$$

For a single diameter pipeline with multiple outlet, factor F or the adjusted factor, $F(\alpha)$, allows rapid calculation of head loss caused by friction. If multi-diameter pipelines are used, the F factors cannot be applied directly to the entire length of the pipeline. For analytical purposes, the pipeline is divided into reaches based on pipeline diameter; then again factor F cannot be applied directly to any, except the most downstream pipe reaches because other reaches of the pipeline would have outflow at their downstream ends.

To resolve this problem, Anwar (1999) proposed a factor G_a to estimate the friction head losses in multi-outlet pipes or pipelines with outflow at the downstream ends. Factor G_a can be applied to each reach within a multi-diameter pipeline to estimate the head loss. An adjusted factor G_a allowed for the first outlet to be at any fraction of outlet spacing from the pipeline inlet. The following assumptions were made in the theoretical development of adjusted factor G_a.

- The outlets are equally spaced and have equal discharge.
- The pipe friction factor remains constant along the length of the pipeline.
- The velocity head can be neglected.
- The increase in pressure past each outlet caused by a decrease in the flow is equal to the head loss caused by turbulence associated with each outlet.
- Head loss at the change in pipe diameter is ignored.

Anwar (1999) proposed the following formula to calculate the G (friction correction factor):

$$G = \frac{1}{N^{m+1}(1+r)^m} \sum_{k=1}^{N}(k+Nr)^m \qquad \text{...(6.17)}$$

where,

N = number of outlets along the pipeline

r = ratio of the outflow discharge to the total discharge through the outlets ($r \geq 0$)

k = integer representing the successive segments of the pipeline.

m and n = exponents of the average flow velocity in the pipeline and internal pipeline diameter, respectively

Further, another equation was proposed for G_a (adjusted friction correction factor for pipelines with multiple diameter equally spaced outlets and outflow).

$$G_a = \frac{N.G + x - 1}{N + x - 1} \qquad \qquad ...(6.18)$$

In this equation, if the first outlet is located at a full outlet spacing from the pipeline inlet, i.e. $x = 1$, then $G_{a \,(for\, x = 1)} = G$

Example 6.4. In a mango orchard, a 150 m long sub-mainline of 50 mm diameter is serving 30 numbers of laterals. The sub-mainline is carrying a discharge of 1.4 lps. Take $C = 140$, $m = 1.852$ and $K = 1.22 \times 10^{12}$. Determine the frictional head loss in the sub-main pipeline by using Hazen-Williams formula and Watters and Keller formula.

Sol.: Hazen-Williams formula can be given as below.

$$h_f = 1.212 \times 10^{12} . \frac{(Q/C)^{1.852}}{D^{4.87}} . \frac{L}{100}$$

where,

h_f = frictional head loss in the pipeline (m)

Q = discharge in the pipeline (lps)

C = friction coefficient to consider pipe material

L = length of the pipeline (m)

D = inside diameter of the pipeline (mm)

$$h_f = 1.212 \times 10^{12} . \frac{(1.4/140)^{1.852}}{50^{4.87}} . \frac{150}{100}$$

$$= 1.91 \text{ m}$$

Now, since there are 30 numbers of outlets are being served with this sub-main pipe, hence a correction factor, F, will also be determined as follows.

$$F = \frac{1}{b+1} + \frac{1}{2N} + \frac{(b-1)^{0.5}}{6N^2}$$

$$F = \frac{1}{1.852+1} + \frac{1}{2 \times 30} + \frac{(1.852-1)^{0.5}}{6(30)^2}$$

$$= 0.36$$

Finally, the friction head loss will be $h_f \times F = 1.91 \times 0.36 =$ **0.68 m**

According to Watters and Keller formula, frictional head loss can be calculated as,

$$h_f = 0.465 \frac{L}{D^{4.75}} Q^{1.75}$$

where,

h_f = pipe friction loss (m)

L = pipe length (m)

D = inside pipe diameter (mm)

Q = pipe flow rate (liters/hr)

$$h_f = 0.465 \frac{L}{D^{4.75}} Q^{1.75}$$

$$= 0.465 \frac{150}{50^{4.75}} 5040^{1.75}$$

$$= 1.11 \text{ m}$$

After including factor F, it will be **0.40 m**

Ans. 0.68 m, 0.40 m

Example 6.5. Online drippers are placed at 2 m spacing on a 100 m long 16 mm diameter lateral line which is serving one row of papaya plantation. The rated discharge of each dripper is 4 lph. If the density of flowing water is 1000 kg/m³ and dynamic viscosity of water is 0.862 × 10⁻³ N-s/m² at 26.7°C, determine the reduction factor (F).

Sol.: Following information is given:

Dripper spacing   = 2 m

Lateral length, L = 100 m

Number of outlets, N = 100 /2 = 50

Diameter of pipe, d = 16 mm

Dripper discharge, q = 4 lph = $\dfrac{4 \times 10^{-3}}{3600}$ = 1.11×10⁻⁶ m³/s

Discharge to be carried in one lateral, Q = 1.11×10⁻⁶ × 50 = 5.55×10⁻⁵ m³/s

Density of water,   = 1000 kg/m³

Dynamic viscosity of water, = 0.862 × 10⁻³ N-s/m²

Reduction Coefficient 'F' is given hereunder:

$$F = \frac{1}{b+1} + \frac{1}{2N} + \frac{(b-1)^{0.5}}{6N^2}$$

where,

b is flow exponent in the head loss equation, i.e. 1.852

N is the number of outlet. In this case, the sub-main is having 100 outlet through which laterals extend. Therefore,

$$F = \frac{1}{1.852+1} + \frac{1}{2 \times 50} + \frac{(1.852-1)^{0.5}}{6 \times 50^2} = 0.36$$

Ans. 0.36

Example 6.6. Estimate the head loss and emission uniformity in a 120 m length of 18 mm diameter lateral line on which 2 drippers per plant are spaced at 0.2 m. Assume $x = 0.70$ and $K_d = 0.80$. Inlet pressure is 200 kPa. The crop is carrots.

Sol.: The flow from the dripper is expressed by the equation:

$$q = K_d H^x$$

Let us take operating pressure head, H = 10 m

$$q = 0.80 \times 10^{0.70} = 4.00 \text{ lph}$$

Total flow rate per lateral, $Q = \frac{120}{0.20} \times 4 = 2400$ lph

The velocity of flow of water in the lateral line,

$$v = \frac{Q}{A} = \frac{2400}{1000 \times 3600 \times \pi \times \left(\frac{9}{1000}\right)^2} = 2.61 \text{ m/s}$$

Now, Reynolds number, $R_e = \frac{vD}{\nu} = \frac{2.61 \times \left(\frac{18}{1000}\right)}{1 \times 10^{-6}} = 46980$

Friction factor, f, can be estimated by:

$$f = \frac{0.316}{(R_e)^{\frac{1}{4}}} = \frac{0.316}{(46980)^{\frac{1}{4}}} = 0.021$$

By using Darcy–Weisbach formula,

$$h_f = f.\frac{L}{D}.\frac{v^2}{2g}$$

$$= 0.021.\frac{120}{18}.\frac{2.61^2}{2 \times 9.81} = 0.048$$

Ans. 0.048 m

7

PLANNING AND DESIGN OF DRIP IRRIGATION SYSTEMS

The planning and design of a drip irrigation system is essential to supply the required amount of irrigation water. The daily water requirement of the plant depends on the water that is taken by the plant from the soil and the amount of water evaporates from the soil in the immediate vicinity of the root zone in a day. The plant water intake is affected by the leaf area, stage of growth, climate, soil conditions, etc. The water requirement and irrigation schedule can be determined from the soil or plant indicators-based methods or soil-water budget method, but the simplest and most commonly method is to use pan evaporation data.

The details of methods of potential evapotranspiration and several techniques of irrigation scheduling have been elaborated in Chapters 4 and 5. To apply the required amount of water uniformly to all the plants in the field, it is required to maintain desired hydraulic pressure in the pipe network. The hydraulic design of drip irrigation system includes determining the size of laterals, manifolds, sub-main, main pipeline, selection of appropriate emitter discharge and size of pumping unit. Major steps for the design of a drip irrigation system are:

    (i)   Collection of general information

    (ii) Layout of the drip irrigation system

   (iii) Crop water requirement

   (iv) Hydraulic design of the system.

7.1 Collection of General Information

General information on water source, crops to be grown, topographic condition, type and texture of the soil and climatic data are essential for designing a drip irrigation system. The information on type of soil will help in selection of emitter and its rated discharge rate. Climatic data are required to estimate the potential evapotranspiration, if pan evaporation data is not available.

7.2 Layout of the Drip Irrigation System

The layout of the drip irrigation system depends upon the length and breadth of the field and location of source of water. The decision as to number and spacing of the laterals is governed by the nature of the crop. Each row, two rows or more than two rows can be served by one lateral. Normally, the sub-mainline runs along the slope and laterals are laid across the slope or along the contour lines of the field.

7.3 Crop Water Requirement

The daily crop water requirement is estimated by using the equation

$$V = E_{PAN} \times K_p \times K_c \times A \times W_p \qquad \qquad ...(7.1)$$

Net volume of water to be applied, $V_n = V - R_e \times A \times W_p$ ...(7.2)

Number of daily operating hours of the system =

$$\frac{V_n}{\text{No. of emitters per plant} \times \text{No. of plants} \times \text{Discharge of one emitter}}$$

where,

V = volume of water required (litres)

E_{PAN} = mean pan evaporation for the month (mm/day)

K_c = crop coefficient

K_p = pan factor

A = area to be irrigated (m²)

R_e = effective rainfall (mm)

W_p = percentage wetting

The crop coefficient (K_c) varies with the crop growth stage and season. It is considered for crop productive stage for the design of a drip irrigation system.

7.4 Hydraulic Design of Drip Irrigation System

Flow in drip irrigation line is hydraulically steady, spatially varied pipe flow with decreasing discharge. The total discharge in lateral, sub-main or main is decreasing with respect to the length of the line. The lateral and the sub-main lines are having similar hydraulics characteristics while the mainline is designed based on input pressure, the required pressure and slope of the energy gradient line which will give a total energy higher than that of required at any sub-main for irrigation. The ideal drip irrigation system is one in which all emitters deliver the same volume of water in a given time. Under field conditions, it is difficult to achieve this goal; however, the emitter flow variation in the lateral line can be controlled by the hydraulic design.

The pipes used in drip irrigation system are made of plastics and are considered smooth. The pressure drop due to friction or frictional head loss can be evaluated with the help of the Darcy-Weisbach equation or Hazen-Williams empirical equation. The details of these equations are given in Chapter 6.

Flow carried by each lateral line (Q_1) = Discharge of one emitter × No. of emitters per lateral

Flow carried by each sub-mainline (Q_s) = Q_1 × No. of lateral lines per sub-main

Flow carried by mainline (Q) = Q_s × No. of sub-mainlines

Total head loss due to friction (H_f)

= Friction head loss in mains + Friction head loss in sub-mains + Friction head loss in laterals + Friction head loss in accessories and fittings

Operating pressure head required at the dripper = H_e

Total static head = H_s

Total pumping head $(H) = H_f + H_e + H_s$

The diameter of the main, sub-mains and laterals are chosen based on the hydraulics of pipe flow. The sizes of main, sub-mains, laterals and pumps are estimated using the procedure stated below:

7.4.1 Design of Lateral Pipe

Proper design of drip irrigation system consists in assuring a high uniformity of water application. The lateral lines are the pipes on which the emitters are inserted. They receive the water from sub-mainline and are usually made of LLDPE ranging in diameter from 12 to 20 mm. The emitter discharge is a function of the lateral pressure. Low discharges and low pressure heads in the distribution network allow using of smaller pipes of lower pressure rating which reduces the costs. Since application of water is slow and spreads over a long-time, peak discharges are reduced, thus requiring smaller size pipes and pumps which causes less wear and longer life of network. The irregularity of emitter discharge is essentially due to the pressure variation in laterals, the land slope, and emitter's characteristics. The discharge of an emitter is also influenced by the water temperature and partial or complete plugging of emitters. When the pipe network is installed, it is difficult to change its design and layout of the system. So, it is essential to assure precision of calculations of frictional losses and pipe sizes.

The design of drip irrigation lateral has been the subject of several studies which has been published in peer reviewed journals. We have tried to incorporate the various methods used in designing of laterals. If we go back, it was the graphical method or polyplot, used by Christiansen (1942). This method got obsolete due to availability of computers and Wu and Gitlin (1974) developed a computer model based on the average discharge. Keller and Karmeli (1974) formulated a computational model to calculate the pressure at any point along lateral by testing many values of emitter's exponent. Computations are considerably simplified by assuming that the emitter discharge is constant along the lateral. Mathematical models have been established using the law of continuity and conservation of energy.

Perolt (1977) used an iterative process based on the back step method to converge the solution. Solomon and Keller (1978) tried the calculation based on the piezometric curve. In order to increase the efficiency of design, researchers became interested in the hydraulic analysis of drip irrigation lateral. The finite element method (FEM) is a systematic numerical procedure that has been used to analyze the hydraulics of the lateral pipe network. A finite element computer model was developed by Bralts and Segerlind (1985) to analyze microirrigation sub-main units. The advantage of their technique included minimal computer storage and application to a large microirrigation network.

Bralts and Edwards (1986) used a graphical technique for field evaluation of microirrigation sub-main units and compared the results with calculated data. Yitayew and Warrick (1988) presented an alternative treatment including a spatially variable discharge function as part of the basic solution to drip irrigation lateral design. They expounded two evaluations: An analytical solution, and a Runge-Kutta numerical solution of non-linear differential equations. Drip irrigation system design was further analyzed using the microcomputer program by Bralts et al. (1991).

This program provided the pressure head and flows at each emitter in the system. The program also gave several useful statistics and provided an evaluation of hydraulic design based upon simple statistics and economics criteria. Since the number of laterals in such a system is large, Bralts et al. (1993) proposed a technique for incorporating a virtual node structure, combining multiple emitters and lateral lines into virtual nodes. After developing these nodal equations, the FEM technique was used to numerically solve nodal pressure heads at all emitters. This simplification of the node number reduced the number of equations and was easy to calculate with a personal computer.

Most numerical methods for analyzing drip irrigation systems utilize the back step procedure, an iterative technique to solve for flow rates and pressure heads in a lateral line based on an assumed pressure at the end of the line. However, a drip irrigation network program needs large computer memory, and a long computer calculation time due to the large matrix equations. A mathematical model was also developed for a microcomputer by Hills and Povoa (1993) analyzing hydraulic characteristics of flow in a drip irrigation system.

Bralts et al. (1993) used the finite element method for numerical solution of non-linear second order differential equations. Their articles provide a detailed description of other methods. Kang and Nishiyama (1994, 1996) also used the finite element method to analyse the pressure head and discharge distribution along lateral and sub-main pipe. The golden section search was applied to find the operating pressure heads of lateral corresponding to the required uniformity of water application.

Valiantzas (1998) introduced a simple equation for direct calculation of lateral hydraulics. Computations are based on the assumption of a no uniform emitter outflow profile. Lakhdar and Dalila (2006) presented a computer model based upon the back step procedure and the control volume method to simultaneously solve non-linear algebraic equations. An alternative iteration process was developed which simplified the model to design lateral of drip irrigation system.

7.4.1.1 Hydraulic Analysis of Laterals

Hydraulic analysis of drip irrigation laterals is based on the hydraulics of pipelines with multiple outlets. The successful design is a compromise between the choice of high uniformity and low installation cost. It is important to calculate the pressure distribution and emitter discharge correctly along the lateral. Using equations of energy and mass conservation, the closing between two sections of an elementary control volume ends up in a two non-linear partial differential equations system, associating pressure and velocity. These equations describe the flow in the lateral; their solution is tedious because of interdependence of the discharge and the pressure in a linear relation. The solution of these equations cannot be completely analytic due to the empiric relation of discharge emitters and the energy loss relations.

Numerical approaches solve the problem either backward or forward and can take into consideration the variability in discharge, pressure, diameter, and spacing. Numerical approaches became popular with the development of personal computers. These approaches to solve the hydraulics of drip systems included the use of finite element methods. Solving the hydraulics of drip irrigation lateral pipelines requires solving sets of non-linear equations which are common in the hydraulics of pipe networks.

There are two main approaches in solving these systems of equations. The first approach is a successive linear approximation method in which these equations are linearised using an initial solution. This results in converting the system of non-linear equations into a set of linear equations. Solving such set of linear equations is quite common in the finite element method where symmetric banded matrices are solved efficiently. The results are used as an improved estimate of an initial solution; then a new system of linear equations is formed and solved again. The procedure is continued until convergence. The successive linear approximation approach was implemented to solve the hydraulics of drip irrigation systems (Hathoot et al., 1993; Bralts and Segerlind, 1985; Kang and Nishiyama, 1996a, 1996b).

The second approach is to use the Newton-Raphson method to solve the system of non-linear equations. This method showed a speed of convergence much faster than successive linear approximation where the two methods started from the same initial solution in analyzing a set of small hypothetical drip irrigation systems (Mizyed, 1997). However, the biggest disadvantage of applying both successive linear approximation and Newton-Raphson methods to drip irrigation systems is requirements of memory. This resulted when all the laterals and outlets in a real size drip irrigation systems were considered.

Besides Finite Element Method and the Newton-Raphson method to solve the sets of linear equations, control volume method and method of Runge-Kutta of order four have also been applied in drip irrigation lateral design. The numeric control volume method (CVM) is often used to determine pressure and discharge in drip irrigation lateral. It is applied to an elementary control volume on the lateral and permits an iterative development, volume after volume, from a lateral extremity to the other. Howell and Hiler (1974) and Helmi et al. (1993) applied this technique to an example of drip irrigation lateral, starting iterative procedure of calculation from the lateral entrance. Thus, knowing the output discharge to the lateral entrance, represented by the sum of average emitters discharge, the trial and error method is successively used till the lateral end, in order to lead to the convergence. However, the risk of obtaining a negative velocity still exists.

This approach seems to provide some precise results but could become slow for numerous reasons of the possible iteration, without excluding the divergence risk. Another method to solve the set of non-linear equations is the numeric method of Runge-Kutta of order four. The Runge-Kutta method allows the integration of the differential equations system of the first order by describing variations of pressure and velocity from the initial conditions to the lateral extremity ($x = 0$). Given the fact that the pressure in this point is unknown, it is therefore necessary to use an iterative process in order to converge toward the solution to the other extremity of the lateral ($x = L$), where the value of the pressure is known (input). The iterative process is assured by the interpolation technique by Lagrange's polynomial.

We have shown here that how the differentials equations are formed and applied to design the laterals. The solution was given by Zella and Kettab (2002). The mathematical model to be derived is a system of two coupled differential equations of the first order. The unknown parameters are pressure and velocity. The principle of mass conservation is first applied to an elemental control volume of length dX the horizontal drip irrigation lateral. This has been shown in Fig 7.1. The discharge entering at the point X will be equal to discharge leaving at the point $X + dX$ and discharge passed through emitter. This can be shown by the following equation.

$$AV_X = AV_{X+dX} + q_e \qquad \qquad ...(7.3)$$

where,

A = cross-sectional area of lateral

V = velocity of flow in the control volume between X and $X + dX$

q_e = emitter discharge which is assumed to be uniformly distributed through the length dX

The emitter discharge expression is given by the following empirical relation.

$$q_e = \alpha H^y \qquad \qquad ...(7.4)$$

where,

α = emitter constant

y = emitter exponent for flow regimes and emitter type

H = pressure at the emitter

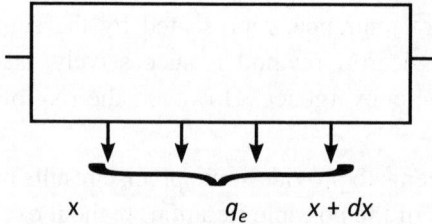

Fig. 7.1: Elemental control volume under consideration

The Bernoulli's equation can be applied to the flowing fluids. The principle of energy conservation is applied to the elemental control volume to give the Bernoulli's equation in the following form:

$$H_X + \frac{1}{2g}V_X^2 = H_{X+dX} + \frac{1}{2g}V_{X+dX}^2 + hf \qquad \qquad ...(7.5)$$

where,

hf = head loss due to friction between X and dX. Its expression can be given in the following form:

$$hf = \alpha V^m dX \qquad \qquad ...(7.6)$$

The Reynolds number is a dimensionless number that gives a measure of the ratio of inertial forces to viscous forces and quantifies the relative importance of these two types of forces for given flow conditions. Regime flow is determined by Reynolds number which can be expressed by the following equation:

$$R_e = \frac{VD}{\mu}$$ ...(7.7)

where,

D = lateral diameter

μ = kinematic viscosity

When $R_e > 2300$, $m = 1.852$ and the value of a in Eqn. 7.6 is given by the following equation using the Hazen-Williams formulation.

$$a = \frac{K}{C^m A^{0.5835}}$$ ...(7.8)

where,

C = Hazen-Williams coefficient

K = coefficient

m = exponent describing flow regime. When $R_e > 2300$, $m = 1$ and the value of a is

$$a = \frac{32\mu}{gD^2}$$ ...(7.9)

where,

g = gravitational acceleration. After expansion of the terms H_{X+dX} and V_{X+dX}, Eqn. 7.5 is written as,

$$H_X + \frac{1}{2g}V_X^2 = H_X + \frac{\partial H_X}{\partial x}dX + \frac{1}{2g}\left(V_X^2 + 2V_X\frac{\partial V_X}{\partial x}dX + \left(\frac{\partial V_X}{\partial x}dX\right)^2\right) + hf$$

...(7.10)

If the term $\left(\frac{\partial V_X}{\partial x}dX\right)^2$ is supposed to be negligible and rearranging the Eqn. 7.10, we get,

$$\frac{\partial H}{\partial x}dX + \frac{V}{g}\frac{\partial V}{\partial x}dX + hf - 0$$ ...(7.11)

By using the expansion of V_{X+DX} in Eqn. 7.3, we get,

$$A\frac{\partial V}{\partial x}dX + q_e = 0$$ ...(7.12)

Finally, by combining Eqns. 7.4, 7.6, 7.9, 7.11 and 7.12, the final system of equations is found as

$$\frac{\partial V}{\partial x} = -\frac{\alpha}{Adv} H^y \qquad \ldots(7.13)$$

and

$$\frac{\partial H}{\partial x} = -aV^m - \frac{V}{g}\frac{\alpha}{Adv} H^y \qquad \ldots(7.14)$$

In order to solve the solution of Eqns. 7.13 and 7.14, the velocity at the end of the lateral $(V_{(X=L)} = 0$ and the pressure head $[H_{(X=0)} = H_{max}]$ are given. These can be integrated by using the method of Runge-Kutta of order 4 by constructing an iteration process. Let us assume that $H_{(L)} = H_{min}$ is known. A new space variable X is defined, such as $X = L - X$. The system of Eqns. 7.13 and 7.14 becomes

$$\frac{\partial V}{\partial X} = \frac{\alpha}{A\Delta x} H^y \qquad \ldots(7.15)$$

$$\frac{\partial H}{\partial X} = aV^m - \frac{\alpha}{Ag\Delta x} VH^y \qquad \ldots(7.16)$$

The initial conditions to this problem are $V_{(X=0)} = 0$ and $H_{(X=0)} = H_{min}$

Iteration Process

To integrate simultaneously Eqns 7.15 and 7.16, we have to provide only two estimates of the pressure head at the downstream end of the lateral $(X = 0)$; this can be written as H^0_{min} and H^1_{min}. Now, two solutions of the initial value problem (7.15) and (7.16) are carried out, yielding H^0_{max} and H^1_{max}. A new estimate of H_{min} can then be obtained by making use of the interpolating Lagrange polynomial of degree one. This new estimate H_{min} is written as follows in order to get the next solution H^2_{max}.

$$H_{min} = \frac{H_{max} - H^1_{max}}{H^0_{max} - H^1_{max}} H^0_{min} + \frac{H_{max} - H^0_{max}}{H^1_{max} - H^0_{min}} H \qquad \ldots(7.17)$$

This process is continued until convergence, which means

$$ErH = \left| \frac{H^{new} - H^{old}}{H^{new}} \right| \prec \varepsilon \qquad \ldots(7.18)$$

or

$$ErV = \left| \frac{V^{new} - V^{old}}{V^{new}} \right| \prec \varepsilon \qquad \qquad ...(7.19)$$

A program of calculation in Fortran can be applied for the two numeric methods and executed on a computer until convergence, i.e. *ErH* and *ErV* to $\varepsilon = 10^{-5}$.

Zella et al. (2006) presented a design methodology for drip irrigation system by using control volume method (CVM) with the back step procedure. The proposed numerical method is simple and consists of delimiting an elementary volume of the lateral equipped with an emitter, called 'control volume' on which the conservation equations of the fluid hydrodynamics were applied. Control volume method is an iterative method which is used to calculate velocity and pressure step-by-step throughout the drip irrigation network based on an assumed pressure at the end of the line. A simple computer program can be developed for the calculation. The use of the control volume method reduced computing time as required in FEM technique and facilitated easier computations.

The proposed CVM model was based upon conservation of mass and energy applied to an elementary control volume which contained one emitter on one lateral or sub-main and solved by the back step procedure. The first control volume was chosen at the end of the last lateral pipe of network to find pressure at the inlet of the lateral pipe (H_{Lmax}), or pressure at the end of lateral pipe (H_{Lmin}). The iterative process based on the back step procedure was successively applied until the other lateral extremity and for the entire network (Fig.7.2). The calculation can be continued step-by-step using an iteration process for all the sub-main units. Figure 7.2 shows the total average flow rate of network (Q_{avg}) in m³/s, which is an input for the computation, the total flow rate Q_T in m³/s given after computation, the total pressure head H_{Tmax} in *m* and the velocity V_{max} in m/s at the inlet of the network.

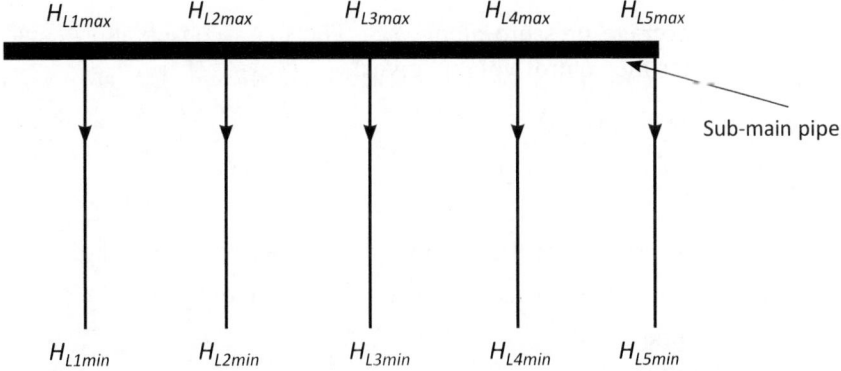

Fig. 7.2: An example of drip irrigation network

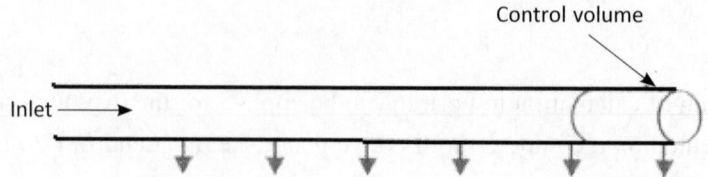

Fig. 7.3: Lateral pipe of drip irrigation showing elementry control volume

The total network is formed by the identical laterals presented in Fig. 7.3. In Fig. 7.3, $H_{L\max}$ represents the pressure at the lateral entrance, $Q_{\max}$ represents the total flow rate at inlet of lateral pipe, $V_{\max}$ the velocity at lateral pipe entrance, $H_{L\min}$ the pressure at the end of lateral pipe, $V_{L\max}$ the velocity at the end of lateral pipe, $Q = q_i$, the discharge of last emitter and L_L the length of lateral pipe. For the elementary control volume (Fig. 7.4), the principles of mass and energy conservation are applied. The *ith* emitter discharge q_i in m³/s was assumed to be uniformly distributed along the length between emitters, Δx_L, and is given by:

$$q_i = \alpha \overline{H}^y \qquad \qquad ...(7.20)$$

or

$$q_i = \alpha \left(\frac{H_i + H_{i+1}}{2} \right)^y \qquad \qquad ...(7.21)$$

where,

α = empirical constant

y = emitter exponent

H_i, H_{i+1} = pressure at *ith* and (i + 1)*th* point

H = average pressure along Δx_L. The mass conservation equation for the control volume gives:

$$M_i \big/ t = M_{i+1} \big/ t + q_i \qquad \qquad ...(7.22)$$

where,

M_i = water mass at the entrance of the control volume, kg

M_{i+1} = water mass at the exit of control volume, kg

t = time in *s*

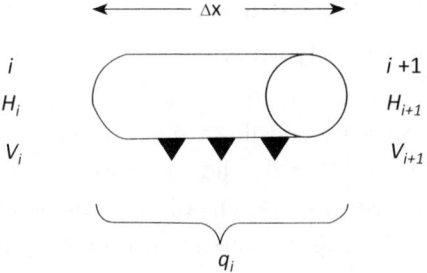

Fig. 7.4: Elementry control volume

The energy conservation between i and $i + 1$ is as follows:

$$E_i = E_{i+1} + \Delta H \qquad \qquad ...(7.23)$$

where,

E_i = flow energy or pressure at the input

E_{i+1} = flow energy at the exit

ΔH = local head loss, hf due to the emitter in m due to friction along Δx_L.
The head losses ΔH are given by the following formula:

$$\Delta H = a \overline{V}_L^m \Delta x_L \qquad \qquad ...(7.24)$$

$$\Delta H = a \left(\frac{V_i + V_{i+1}}{2} \right)^m \Delta x_L \qquad \qquad ...(7.25)$$

V_L in m/s, is the average velocity between i and $(i + 1)$, V_i and V_{i+1} are velocity, respectively at *ith* and $(i + 1)th$ cross-section lateral, the value of parameter α is given by Hazen-Williams equations:

For turbulent flow Reynolds number, $R_e > 2300$,

$$a = \frac{K}{C^m A_L^{0.5835}} \qquad \qquad ....(7.26)$$

For laminar flow, $R_e < 2300$,

$$a = \frac{32v}{gD_L^2} \qquad \qquad ...(7.27)$$

where,

C = Hazen-Williams coefficient

K = proportional coefficient

m = exponent ($m = 1$ for laminar flow, $m = 1.852$ for turbulent flow)

A_L = cross-sectional area of lateral pipe, m^2

D_L = interior lateral pipe diameter in m

v = kinematic viscosity (m²/s)

g = gravitational acceleration (m/s²). H_L and V_L respectively, are the average pressure and the average velocity between *ith* and $(i + 1)th$ emitter on the lateral. The calculation model for lateral pipe solves simultaneously the system of two coupled and non-linear algebraic equations, having two unknown values: V_{i+1} and H_{i+1}.

Equations (7.22) and (7.23) become:

$$A_L V_i = A_L V_{i+1} + q_i \qquad \qquad ...(7.28)$$

$$H_i + \frac{V_i^2}{2g} = H_{i+1} + \frac{V_{i+1}^2}{2g} + a\left(\frac{V_i + V_{i+1}}{2}\right)^m \Delta x \qquad ...(7.29)$$

and equations (7.28) and (7.29) become:

$$V_{i+1} = V_i - \frac{\alpha}{A_L}\left(\frac{H_i + H_{i+1}}{2}\right)^y \qquad \qquad ...(7.30)$$

$$H_{i+1} = H_i + \frac{1}{2g}\left(V_i^2 - V_{i+1}^2\right) - a\left(\frac{V_i + V_{i+1}}{2}\right)^m \Delta x \qquad ...(7.31)$$

For the lateral, equations (7.30) and (7.31) become:

$$\left(\frac{dV}{dx}\right)_L = -\alpha \frac{\overline{H}_L^y}{A_L \Delta x_L} \qquad \qquad ...(7.32)$$

$$\left(\frac{dH}{dx}\right)_L = -a\overline{V}_L^m \qquad \qquad ...(7.33)$$

For sub-main pipe, equations system is

$$\left(\frac{dV}{dx}\right)_S = \frac{Q_S}{A_S \Delta x_S} \qquad \qquad ...(7.34)$$

$$\left(\frac{dH}{dx}\right)_S = -a\overline{V}_S^m \qquad \qquad ...(7.35)$$

where,

Q_s = flow rate in sub-main pipe

A_s = cross-sectional area of sub-main pipe

V_s and H_s, respectively, are velocity and pressure in sub-main pipe. At the end of the lateral $V_{i=0}$, H_{Lmax} is inlet head pressure given at entrance of lateral pipe. The slope of lateral and sub-main pipe is assumed null. When H_{Lmax} is fixed, the computation program of lateral can give the distribution velocity or emitter's discharge and pressure along lateral. Theoretical development giving equations (7.30), (7.31) and (7.32), (7.33) can be solved without the use of matrix algebra through CVM.

As we know, the economics of drip irrigation largely depends upon size and length of drip irrigation lateral. Since drip irrigation system requires large amount of pipe per unit of land, the pipe cost must be economically feasible. Appropriate design saves the cost and ensure reliability of the drip system. Besides, modifying the conventional crop geometry can also considerably reduce drip irrigation system cost. The plant-to-plant and row-to-row spacing can be changed without changing the plant population per unit area. The design of drip irrigation lateral is mainly concerned with the selection of the pipe size for a given length which can supply the estimated amount of water to the plants keeping the desired range of uniformity. Drip irrigation lateral design is classified into mainly two types of design problems:

1. Lateral length is unknown and pipe size is given;
2. Pipe size is unknown but lateral length is constrained;

Usually, pipe sizes are limited to pipe diameter less than 20 mm by economics. The information required for designing laterals are field slope, emitter flow rate, number of emitters per plants, emitter flow function, plant spacing and desired uniformity. In first type of problem, the maximum lateral length is determined while maintaining the required uniformity value. The allowable energy loss can be determined using Fig. 7.5 and the following relationships:

$$\frac{P_o}{P_n} = \left[\frac{q_o}{q_n}\right]^{1/x} \qquad \qquad ...(7.36)$$

$$P_o = \left(\frac{P_o}{P_n}\right) \times (\gamma) \times \left[\frac{q_r}{K_e}(2 - \frac{q_o}{q_n})\right]^{1/x} \qquad ...(7.37)$$

$$H_n = \frac{P_n}{\gamma} ; H_o = \frac{P_o}{\gamma} \qquad \qquad ...(7.38)$$

$$\Delta H = H_n - H_o + S_o L \qquad \qquad ...(7.39)$$

$$L = 1.70 \, \Delta H^{0.35} \, D^{1.71} \left(\frac{C S_e}{100 q_r}\right)^{0.65} \qquad ...(7.40)$$

where,

P_n = pressure at the head of the lateral (kPa)

P_o = pressure at the end of the lateral (kPa)

$q_n = q_o$ = emitter flow rate at the head and end of the lateral, respectively (lph)

γ = specific weight of water (9.81 kN/m³)

H_n and H_o = pressure head at the inlet and outlet of the lateral, respectively (m)

ΔH = pipe friction energy loss (m)

D = inside pipe diameter (mm)

S_e = emitter spacing (m)

C = Hazen-Williams roughness coefficient

S_o = field slope (positive for down and negative for up)

L = lateral length (m)

K_e = proportionality factor that characterizes the emitter dimensions

For the second type of problem the following formula can be used for calculating pipe diameter.

$$D = 0.73 \, (\Delta H)^{-0.20} \, (L)^{0.58} \left(\frac{100 q_r}{C S_e} \right)^{1.852} \qquad ...(7.41)$$

When water is removed through drippers along the laterals, the friction loss for the given diameter and length of lateral will be less than if the flow was constant for the entire length. A reduction factor (F) is multiplied with the estimated frictional loss which can be estimated by

$$F = \frac{1}{m+1} + \frac{1}{2N} + \frac{\sqrt{m-1}}{6N^2} \qquad ...(7.42)$$

where,

m = 1.852 for the Hazen-Williams equation and 2 for the Darcy-Weisbach equation, and

N = number of outlets on the lateral.

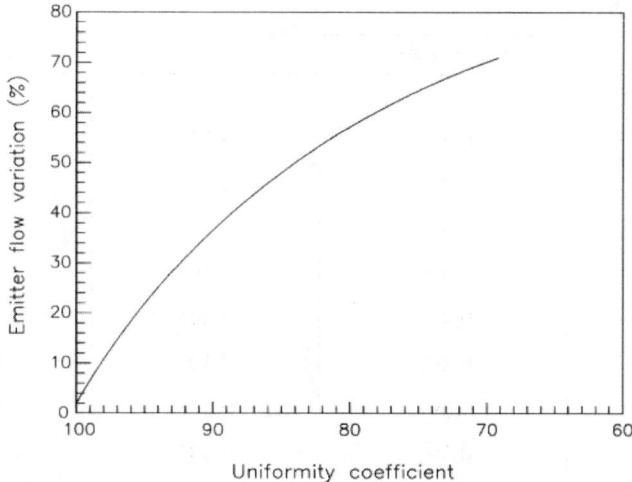

Fig. 7.5: Relationship between emitter flow variation and uniformity coefficient

The values of F for different 'm' are given in Table 7.1. The emitter discharge decreases with respect to the lateral length when the lateral length is laid on zero slope or uphill. When the lateral pipe is laid on mild downhill slopes, the emitter discharge decreases with respect to the lateral length and reaches a minimum emitter discharge and then increases with respect to the length of the lateral line. This is because, the gain of energy due to the land slope at a downstream section is larger than the energy drop by friction.

There is yet another situation where the emitter discharge increases with respect to the length of lateral line. This is caused by steep slopes where the energy gained by the slopes is larger than friction drop for all section along the lateral line. Use of series of pipe sizes in laterals or sub-main design will help to reduce the maximum pressure variation. By changing the pipe size, it is possible to make the friction drop approach more closely to the energy gain at all points along the line. The line slope of each section can be used as the energy slope to design the size of the lateral and sub-mains. This can be used for both uniform and non-uniform slopes.

Table 7.1: Values of F for different numbers of outlets

Number of outlets	F		
	m = 1.85	*m = 1.9*	*F = 2.0*
1	1.000	1.000	1.000
2	0.639	0.634	0.625
3	0.535	0.528	0.518
4	0.486	0.480	0.469

Contd.

Contd.

Number of outlets	F		
	m = 1.85	*m = 1.9*	*F = 2.0*
5	0.457	0.451	0.440
6	0.435	0.433	0.421
7	0.425	0.419	0.408
8	0.415	0.410	0.398
9	0.409	0.402	0.391
10	0.402	0.396	0.385
11	0.397	0.392	0.380
12	0.394	0.388	0.376
13	0.391	0.381	0.373
14	0.387	0.381	0.370
15	0.384	0.379	0.376
16	0.382	0.377	0.365
17	0.380	0.375	0.363
18	0.379	0.373	0.361
19	0.377	0.372	0.360
20	0.376	0.370	0.359
22	0.374	0.368	0.357
24	0.372	0.366	0.355
26	0.370	0.364	0.353
28	0.369	0.363	0.351
30	0.368	0.362	0.350
35	0.365	0.359	0.347
40	0.364	0.357	0.345
50	0.361	0.355	0.343
100	0.356	0.350	0.338
more than 100	0.351	0.345	0.333

Example 7.1. Determine the length for a 16 mm inside diameter drip irrigation lateral laid out in a row of young mango plants, spaced 6 m apart with 4 emitters supplying water to each plant. The rated discharge capacity of emitter is 4 lph with emitter flow function, $q = 2.52 \, (P/\gamma)^{0.8}$. The slope of the field is 1.0% and uniformity coefficient of 95% is desired to be achieved.

Sol.: As the emitter flow function is given as $q = 0.63\ (P/\gamma)^{\,0.8}$, therefore, equating it with Eqn. 3.5., we get $K_e = 2.52$ and $x = 0.8$.

Since 4 emitters are being used for one plant, the required emitter flow discharge is

$$q_r = 4 \times 4 = 16 \text{ lph}$$

For uniformity coefficient of 95%, q_n/q_o is estimated as 1.2 from Fig. 7.5.

$$\frac{P_o}{P_n} = \left[\frac{q_o}{q_n}\right]^{1/x} = \left(\frac{1}{1.2}\right)^{1/0.8} = 0.86$$

Pressure at the end of the lateral can be determined by Eqn. 7.37.

$$P_o = (0.86) \times (9.81) \times \left[\frac{16}{2.52}\ (2 - \frac{1}{1.2})\right]^{1.25} = 96 \text{ kPa}$$

Pressure at the head of the lateral by using Eqn. 7.36,

$$P_n = P_o \times (1.2)^{1/x} = (96) \times (1.2)^{1.25} = 120 \text{ kPa}$$

Pressure head at the head of the lateral,

$$H_n = \frac{P_n}{\gamma} = \frac{120}{9.81} - 12.2 \text{ m}$$

Pressure head at the end of the lateral,

$$H_o = \frac{P_o}{\gamma} = \frac{96}{9.81} = 9.7 \text{ m}$$

Now, it is given that $D = 16$ mm; $S_e = 6$ m; $q = 16$ lph; $C = 130$
Pipe friction energy loss will be determined by using Eqn. 7.39.

$$\Delta H = H_n - H_o + S_o L$$

$$= 12.2 - 9.7 + (0.01)\ (L)$$

$$= 2.5 + 0.01\ L$$

Now, putting the value of ΔH in the Eqn. 7.40, we get the following equation to determine the length of the lateral as follows:

$$L = 1.70\ (\ 2.5 + 0.01L\ )^{0.35}\ (16\ )^{1.71} \left(\frac{130 \times 6}{100 \times 16}\right)^{0.65}$$

The length of the lateral can be determined by trial and error approach in MS Excel. Calculate all the terms separately and then multiply each term and change the value of L. After few trials, $L = 208$ m for a given pipe size of 16 mm at the desired uniformity coefficient of 95% was obtained, however this pipe length is too long. By reducing the diameter, pipe length can be reduced.

7.4.2 Design of Sub-main

The sub-main line hydraulics is similar to the principles applied in the design of lateral hydraulics. The sub-main hydraulics characteristics can be computed by assuming the laterals are analogous to emitters on lateral line. Hydraulics characteristics of sub-main and mainline pipe are usually taken hydraulically smooth since PVC and HDPE pipe are normally used. The Hazen-Williams roughness coefficients (C) range from 130 to 150. The energy loss in the sub-main can be computed with the methods similar to those used for lateral computations.

7.4.3 Design of Main

Usually, the pressure controls or adjustments are provided at the sub-main inlet. Therefore, energy losses in the mainline should not affect system uniformity. There is no outlet in case of mainline; therefore reduction factor is not multiplied. The frictional head loss in main pipeline is calculated in the same way using Hazen-Williams, Darcy-Weibach or Watters and Keller's formula.

7.5 Pump Horse Power Requirement

Total dynamic head (H) is determined by adding all the frictional head losses through lateral, sub-main and mainline, total static head and operating pressure required at the emitters. The horse power of pump is estimated by using the estimated dynamic head and pump discharge.

$$\text{Pump horse power } (hP) = \frac{H \times Q}{75 \times \eta_p \times \eta_m} \qquad ...(7.43)$$

where,

H = total dynamic head (m)

Q = total discharge through main line (lps)

η_p and η_m = efficiency of pump and motor, respectively

Example 7.2. The drip irrigation system is to be designed for 1 ha area where mango plantation of 5 years age is done with a plant-to-plant and row-to-row spacing of 5 m × 5 m. The percentage of wetted area and crop coefficient are considered as 50% and 0.80, respectively. The maximum daily peak pan evaporation and pan coefficient were found to be 10 mm and 0.70, respectively. The pumping capacity 2.18 lps and the total dynamic head is 25 m. Take the pump efficiency as 75%. Determine the daily water requirement of mango and the horse power requirement of the pump.

Sol.: The following information is given:

Area, A = 1 ha; spacing = 5 × 5 m; crop coefficient = 0.80; percentage wetted area, WA = 0.50; pan coefficient = 0.70; daily open pan evaporation = 10 mm; $Q = 2.18 \times 10^{-3}$ m^3/s; total dynamic head, H = 25 m and pump efficiency, $\eta_p = 0.75$.

Daily crop water requirement = Crop coefficient × Pan coefficient × Percentage of wetted area × Daily pan evaporation × Area to be irrigated

$$= 0.80 \times 0.70 \times 0.50 \times \frac{10}{1000} \times 10,000 = 28 \text{ m}^3$$

Pump horse power, $h_p = \dfrac{H \times Q}{75 \times \eta_p} = \dfrac{25 \times 2.18}{75 \times 0.75} = 0.971 \approx 1 \text{ hp}$

Ans. 28 m^3/day, 1 hp.

Example 7.3. A citrus plantation has 5 m × 5 m spacing and 40% of the plant area is wetted area with one lateral serving each row with a dripper of 4 lph capacity. The plant root zone depth is 1.2 m and hydraulic conductivity of the soil is 5.65×10^{-6} m/s. The daily peak reference evapotranspiration was observed to be 7 mm. If the emission uniformity is 90%, determine the number of drippers required for each plant.

Sol. We need to find the wetted diameter of one dripper. This will help in determining the number of drippers per plant. The equation for determining the wetted width proposed by Schwartzman and Zur (1985) will be used which is given as hereunder:

where,

w = wetted width or diameter of wetted soil (m)

z = depth of wetting front or root zone depth of crop (m)

q = discharge of emitter (lph)

K = hydraulic conductivity of soil m/s

$$w = 0.0094(1.2)^{0.35}(4)^{0.33}(5.65 \times 10^{-6})^{-0.33} = 0.85 \text{ m}$$

Wetted area by each dripper = $\dfrac{\pi}{4} D^2 = \dfrac{\pi}{4} 0.85^2 = 0.57 \text{m}^2$

Area covered by each plant = 5×5 = 25 m^2

If we take 40% area as wetted area, it becomes $2.70 \times \dfrac{40}{100} = 1.08$ m^2

So, this much area needs to be wetted by a dripper which can wet an area of 0.57 m². Therefore, number of drippers required for each plant $= \dfrac{1.08}{0.57} = 1.89 \approx 2$

Ans. 2

Example 7.4. The rated discharge of a dripper is 8 litre/hr. The spacing between the drippers is 1 m and the lateral spacing is 2 m. Determine the application rate of the dripper in mm/hr.

Sol.: Application rate of the dripper

$$\frac{8L/\text{hr}}{1\,\text{m} \times 2\,\text{m}} = \frac{8 \times 1000}{1000 \times 1 \times 2} = 4\,\text{mm/hr}$$

Ans. 4 mm / hr

Example 7.5. Banana plantations were planned in an area of 5 ha under drip irrigation system. Source of water is a tube-well and located at top left corner of the field. Banana was planted at spacing of 2 m × 2 m in sandy loam soil. Infiltration rate of the soil was measured as 9 mm/day. Peak reference evapotranspiration was determined as 10 mm/day. The root zone depth is 90 cm. The field capacity and permanent wilting point were 30 and 12%, respectively. Density of the soil is 1.42 g/cm³ and application efficiency is 90%. If irrigation is applied at 20% moisture depletion, determine the irrigation interval in days.

Sol.: Depth of irrigation to be applied in one irrigation (net depth of irrigation) = (FC – PWP) × Density × Root zone depth × Percentage of moisture depleted ×10

= (30 – 12) × 1.42 × 0.90 × 0.20 × 10

= 46 mm

Gross depth of irrigation $= \dfrac{46}{0.90} = 51\,\text{mm}$

Irrigation interval $= \dfrac{\text{Gross depth of irrigation}}{\text{Peak reference evapotranspiration}} = \dfrac{51}{10} = 5.1 \approx 5\ \text{days}$

Ans. 5 days

Example 7.6. Design a drip irrigation system for a mango orchard of 1 ha area with length and breadth of 100 m each. Mango plants have been planted at a spacing of 5 m × 5.5 m and the age of crop is 3 years. The maximum pan evaporation during summer is 12 mm/day. The other relevant data are given below:

Land slope   = 0.40 % upward

Water source = A well located at the S–W corner of the field

Soil texture = Sandy loam
Field capacity (FC) = 16 %
Wilting point (WP) = 8 %
Apparent specific gravity (AS) = 1.4 g/cc
Effective root zone depth (Zr) = 120 cm
Maximum allowable deficit (MAD) = 20%
Wetting area percentage (WA) = 30 %
Pan coefficient = 0.7
Crop coefficient = 0.8

Sol.:

Step 1. Estimation of Water Requirement

Evapotranspiration of the crop = Open pan evaporation ×

$$\text{Pan coefficient} \times \text{Crop coefficient}$$
$$= 12 \times 0.7 \times 0.8$$
$$= 6.72 \text{ mm/day}$$

Net depth of water application = (FC – WP) × AS × Zr × 10 × MAD × WA
$$= (16 - 8) \times 1.4 \times 1.20 \times 10 \times 0.20 \times 0.30$$
$$= 8.06 \text{ mm}$$

If the efficiency of the system is 95%, $E_s = 0.95$

Gross depth of water application $= \dfrac{8.06}{0.95} = 8.48$ mm

Step 2. Emitter Selection and Irrigation Time

Emitters are selected based on the soil moisture movement and crop type. Assuming three emitters of 4 lph, placed near each plant in a triangular pattern are sufficient to wet the effective root zone of the crop.

Total discharge delivered in one hour = 4 × 3 = 12 lph

Maximum crop evapotranspiration is already known as 6.72 mm/day.

Irrigation interval $= \dfrac{8.48}{6.72} = 1.26$ days ≈ 30 hrs

Application rate of the system $= \dfrac{q}{(S_e \times S_l)WA}$

$$= \dfrac{12}{5.0 \times 5.5 \times 0.30}$$
$$= 1.45 \text{ mm/hr}$$

Duration of water application $= \dfrac{\text{Gross depth of water application}}{\text{Application rate of the system}}$

$$= \frac{8.48}{1.45}$$

$$= 5.8 \text{ hrs}$$

Step 3. Discharge Through Each Lateral

A well is located at one corner of the field. Sub-main will be laid from the centre of field (Fig. 7.6). Therefore, the length of main, sub-main, and lateral will be 50 m, 97.25 m, 47.5 m each, respectively. The laterals will extend on both sides of the sub-main. Each lateral will supply water to 10 mango plants.

Total number of laterals $= (100/5.5) \times 2$

$= 36.36$ (considering only 36)

Discharge carried by each lateral, $Q_{lateral} = 10 \times 3 \times 4 = 120$ lph

Total discharge carried by 36 laterals $= 120 \times 36 = 4320$ lph

Each plant is provided with three emitters, therefore, total number of emitters will be $36 \times 10 \times 3 = 1080$

Step 4. Determination of Manifolds

Assuming the pump discharge = 2.5 lps = 9000 lph

Number of laterals that can be operated by each manifold = 9000/120 = 75

So, only one manifold or sub-main can supply water to all the laterals at a time.

Step 5. Size of Lateral

Once the discharge carried by each lateral is known, then size of the lateral can be determined by using the Hazen-Williams equation

$$h_f(100) = K \times \frac{(Q/C)^{1.852}}{D^{4.87}} \times \frac{L}{100} \times F$$

$$F = \frac{1}{m+1} + \frac{1}{2N} + \frac{\sqrt{m-1}}{6N^2}$$

where,

$H_f(100)$ = a friction loss per 100 m of pipe (m/100 m)

C = a coefficient of retardation based on the type of pipe material

Q = flow of water in the line (lps)

D = inside diameter of pipe (mm)

K = a constant, 1.22×10^{12} for metric units

F = reduction factor, depends upon the number of outlets

$m = 1.852$

N = number of outlets on the lateral

L = length of pipe (m)

$$F = \frac{1}{1.852+1} + \frac{1}{2 \times 30} + \frac{\sqrt{1.852-1}}{6(30)^2}$$

$$= 0.367$$

$$h_f(100) = 1.22 \times 10^{12} \frac{(0.033/130)^{1.852}}{(12)^{4.871}} \times \frac{47.5}{100} \times 0.367 = 0.26 \text{ m}$$

We also calculated h_f for D = 16 mm and it is 0.063 m. The permissible head loss due to friction is 10% of head of 10 m (head required to operate 4 lph emitters) is 1m, therefore 12 mm diameter lateral pipe size is selected. You can select 16 mm pipe also but the cost will be more.

Step 6. Size of Sub-main

Total discharge through the sub-main = $Q_{lateral}$ × Number of laterals
$$= 120 \times 36$$
$$= 4320 \text{ lph} = 1.2 \text{ lps}$$

Assuming the diameter of the sub-main as 50 mm. The values of parameter of the Hazen-Williams equation are

$C = 150$, $Q = 1.2$ lps, $D = 50$ mm, $K = 1.22 \times 10^{12}$, $F = 0.364$

$$h_f(100) = 1.22 \times 10^{12} \times \frac{(1.2/150)^{1.852}}{50^{4.87}} \times 0.364$$

$$= 0.31 \text{ m}$$

h_f for 97.25 m of sub-main pipe length = 0.31 × (97.25/100) = 0.30 m

Therefore, frictional head loss in the sub-main = 0.30 m

Head at the inlet of the sub-main = $H_{emitter} + h_{f\,lateral} + h_{f\,sub-main} + h_{slope}$
$$= 10 + 0.26 + 0.30 + 0.40 = 10.96 \text{ m}$$

Pressure head variation $= \dfrac{10.96-10.26}{10.96} \times 100$

$$= 6.38 \%$$

Estimated head loss due to friction in the sub-main is much less than the recommended 20% variation, hence reducing the pipe size from 50 to 35 mm will probably be a good option.

$$h_f\ (100) = 1.22 \times 10^{12} \times \frac{(1.2/150)^{1.852}}{35^{4.87}} \times 0.364$$

$$= 1.75\ \text{m}$$

h_f for 97.25 m pipe = 1.75 × (97.25/100) = 1.70 m

Head at the inlet of the sub-main = $H_{\text{emitter}} + h_{f\,\text{lateral}} + h_{f\,\text{sub-main}} + h_{\text{slope}}$

$$= 10 + 0.26 + 1.70 + 0.40 = 12.36\ \text{m}$$

Pressure head variation $= \dfrac{12.36 - 10.26}{12.36} \times 100$

$$= 17\%$$

Pressure head variation lies within the acceptable limit, hence accepted.

Step 7. Size of the Main Line

Assuming the diameter of main pipe as 50 mm

Discharge of main, Q_{main} = Discharge of sub-main, Q_{submain}

The values of parameter of the Hazen-Williams equation are

$$C = 150,\ Q = 1.2\ \text{lps},\ D = 50\ \text{mm},\ K = 1.22 \times 10^{12},$$

$$h_f\ (100) = 1.22 \times 10^{12}\ \frac{(1.2/150)^{1.852}}{(50)^{4.871}}$$

$$= 0.84\ \text{meter}$$

h_f for 50 m main pipe = 0.84 × (50/100) = 0.42 m

Step 8. Determining the Horse Power of Pump

Assume head variation due to uneven field variations and the losses due to pump fittings, etc. as 10 % of all other losses.

$$H_{\text{local}} = 10\ \%\ \text{of all other loss}$$

Total dynamic head = ($H_{\text{emitter}} + h_{f\,\text{lateral}} + h_{f\,\text{sub-main}} + h_{f\,\text{main}} + h_{\text{slope}}$)

$$+ H_{\text{static}} + H_{\text{local}}$$

$$= 12.36 + 0.42 + 10 + 1.28$$

$$= 24.06\ \text{m}$$

Pump horse power (hp) = $\dfrac{H \times Q}{75 \times \eta_p}$

where,

    H = total dynamic head (m)

    Q = total discharge through mainline (lps)

    η_p = efficiency of pump

$$hp = \frac{1.2 \times 24.06}{75 \times 0.60} = 0.64 \approx 1.0$$

Hence, 1 hp pump is adequate for operating the drip irrigation system to irrigate for 1 ha area of mango plantation (Fig. 7.6).

The design details are given as

Operating time of the system	= 5.8 hrs
Irrigation interval	= 30 hrs
Length of laterals	= 47.5 m
Number of laterals	= 36
Diameter of lateral	= 12 mm
Length of sub-main	= 97.25 m
Number of sub-main	= 1
Diameter of sub-main	= 35 mm
Length of main	= 50 m
Number of main	= 1
Diameter of main	= 50 mm
Total power required	= 1 hp

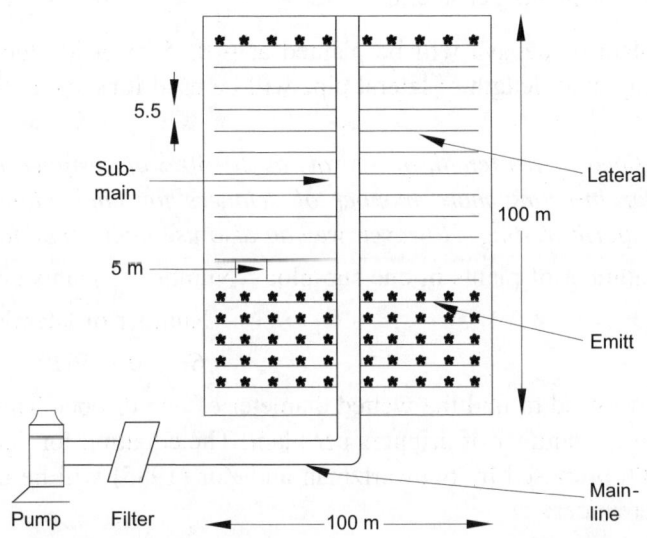

Fig. 7.6: Layout of drip irrigation system for 1 ha of mango orchard

Example 7.7. Determine the discharge capacity of drip irrigation system for papaya plantation on a nearly flat land with sandy loam soil. The land is having 504 m length and 100 m width. The source of water is located at the top edge corner of the field. The plant to plant spacing is 1.5 m and row-to-row spacing is 1.8 m. The daily evaporation rate observed with a class A pan is 10 mm. Take root zone depth of papaya as 1.2 m, hydraulic conductivity of soil as 5.65×10^{-6} m/s, Crop coefficient as 0.95, pan coefficient as 0.70 and percentage of wetted area as 80%. Irrigation interval is 4 days with duration of each irrigation as 2 hrs/day.

Sol.:

Let us divide the whole field into 4 parts and each will be having 126 m length and 100 m width. Each segment or sub-plot is proposed to be irrigated sequentially (Fig. 7.7).

Total area of the field = 504 ×100 = 50400 m^2 = 5.04 ha (1 ha = 10,000 m^2)

As the row to row spacing is 1.8 m, therefore, the laterals will also be spaced at 1.8 m. (*However, in case of close growing crops like vegetables, one lateral can serve two rows or sometimes 4 rows.*)

Number of laterals in each sub-plot $= \dfrac{126}{1.8} = 70$

Length of lateral in each sub-plot = 100 m

Total length of laterals for entire field = Number of laterals in each sub-plot × Nos. of sub-plot × Length of lateral for each row

= 70 × 4 × 100 = 28,000 m

Number of plants per lateral $= \dfrac{100}{1.5} = 66.66 \approx 66$

Last plant of papaya will be planted at = 0.75 m inside the boundary of the field. Any extra length of lateral pipe will be used for closing the laterals by bend clip.

(*Sometimes, extra length of laterals as 20–30% of designed lateral length are used for inserting more number of drippers for each plant in order to maintain uniform wetting. However, we can also use micro-tube for the same*).

Total number of plants in one sub-plot = Number of plants per lateral

× Number of laterals

= 66 × 70 = 4620

Now, we need to find the wetted diameter of one dripper. This will help in determining the number of drippers per plant. The equation for determining the wetted width proposed by Schwartzman and Zur (1985) will be used which is given as hereunder:

$$w = 0.0094 \ (z)^{0.35} \ (q)^{0.33} \ (K)^{-0.33}$$

where,

w = wetted width or diameter of wetted soil (m)

z = depth of wetting front or root zone depth of crop (m)

q = discharge of emitter (lph)

K = root zone depth and hydraulic conductivity of soil as 5.65×10^{-6} m/s

$w = 0.0094(1.2)^{0.35} \, 4^{0.33} \, (5.65 \times 10^{-6}) - 0.33$

   $= 0.85$ m

Wetted area by each dripper $= \dfrac{\pi}{4}D^2 = \dfrac{\pi}{4}0.85^2 = 0.57\text{m}^2$

Area covered by each plant $= 1.5 \times 1.8 = 2.70$ m^2

If we take 80% area as wetted area, it becomes = 2.16 m^2

So, this much area needs to be wetted by a dripper which can wet an area of 0.57 m^2. Therefore, number of dripper required for each plant

$$= \frac{2.16}{0.57} = 3.78 \approx 4$$

There, total number of drippers in each sub-plot = Number of plants in

each sub-plot × 4

= 4620 × 4 = 18480

Water distribution efficiency $\qquad\qquad$ = 95%

Volume of water to be applied daily in one sub-plot = Crop coefficient × Pan coefficient × Percentage of wetted area × Daily pan evaporation × Area to be irrigated

$$= 0.95 \times 0.70 \times 0.80 \times \frac{10}{1000}\text{m} \times (126 \times 100 \text{ m})$$

$$= 67.032 \text{ m}^3$$

$$= 67032 \text{ litres}$$

Gross volume of water to be applied daily in each sub-plot

$$= \frac{67032}{0.95} = 70560 \text{ litres}$$

Irrigation interval $\qquad$ = 4 days and assume the water application

efficiency to be 90 %

Duration of each irrigation = 2 hrs/day

Capacity of drip irrigation system =

$$\frac{\text{Volume of water to be applied daily}}{\text{Duration of irrigation} \times \text{Application efficiency}} = \frac{70560}{2 \times 0.90} = 39200$$

Therefore, the capacity of the proposed drip irrigation system is **10.88 lps.**

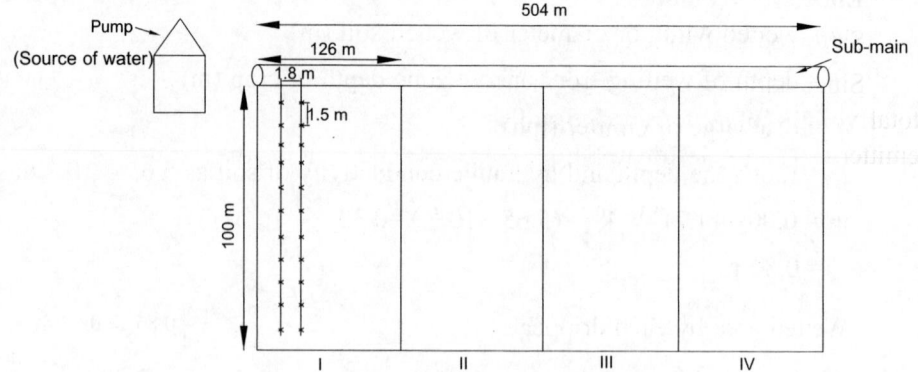

Fig. 7.7: Layout of drip irrigation system

Ans. 10.88 lps.

Example 7.8. The field is located in old alluvial zone and characterized by moderate cold. The average annual rainfall is 1500 mm. Winter season, i.e. November to February receives scanty rainfall. The area of the field is 21.45 ha with majority of sandy loam soil. The field capacity of the soil is 16% and permanent wilting point is 6%. The infiltration rate of the soil is 14 mm/hr. The planting distance of tomato crop is 1.8 m × 0.4 m. Determine the required horsepower for the system. Management allowable deficit is 20%. Determine the horsepower of pump to operate an efficient drip irrigation system. Assume the suitable values, if required.

Sol.:

Since planting spacing is 1.8 × 0.4 m,

$$\text{Plant population per ha} = \frac{10000}{1.8 \times 0.4}$$

$$= 13889$$

Effective crop root depth is assumed as 0.60 m. We have crop coefficient (K_c) values for different growth stage of tomato. The K_c value for highly evaporative demand period, i.e. mid-season (70–115 day) will be considered as 0.80. The pan evaporation is 10 mm and pan coefficient is taken as 0.84.

$$\text{Maximum crop evapotranspiration} = E_{pan} \times K_c \times K_p$$

$$= 10.0 \times 0.80 \times 0.84$$

$$= 6.72 \text{ mm/day}$$

Discharge rate = 4 lph

Considering one emitter per plant,

Distance between lateral = 1.8 m

Distance between emitter = 0.4 m

Emitter selection:

Saturated hydraulic conductivity of soil = 1.8 cm/hr or 5×10^{-6} m/s

Since duration of water application is 1.48 hrs (calculated later in this part), total volume of water applied will be $4 \times 1.48 = 5.92$l. Wetted width of one emitter (w) can be estimated Eqn. 2.3.

$$w = 0.031(V_w)^{0.22} \left(\frac{K}{q} \right)^{-0.17}$$

$$= 0.031(5.92)^{0.22} \left(\frac{5 \times 10^{-6}}{4} \right)^{-0.17}$$

$$= 0.46 \text{ m} = 46 \text{ cm}$$

Therefore, the radius of wetted circle will be 23 cm.

We can also estimate the maximum radius of wetted circle with the following expression.

$R_{MAX} = \sqrt{\dfrac{q}{\pi \times k}}$ (This comes from $q^2 = \pi R_{MAX}^2 \cdot k$ and we know that $q =$AV, here V is replaced by hydraulic conductivity. This needs to be verified in field conditions.)

where, q is cm^3/hr and saturated hydraulic conductivity, K, is in cm/hr.

$$R_{MAX} = \sqrt{\frac{4000}{3.14 \times 1.8}}$$

$$= 26.60 \text{ cm}$$

Therefore, wetted width = 53.20 cm

Wetted area of one emitter = πR_{MAX}^2

$$= 3.14 \times (26.60)^2$$

$$= 2221.73 \text{ cm}^2$$

$$= 0.22 \text{ m}^2$$

Length of the field = 715 m

Width of the field = 300 m

Total number of lateral at a distance of 1.8 m

$$= \frac{\text{Width of field}}{\text{Spacing of laterals}}$$

$$= \frac{300}{1.80} = 166.66 \approx 168 \quad \text{even number}$$

$$\text{No. of emitters per lateral} = \frac{\text{Length of field}}{\text{Spacing of emitter}}$$

$$= \frac{715}{0.4}$$

$$= 1787.5 \approx 1788$$

Here, it is assumed that length of a single lateral is length of field.

Total no. of emitters in the field = No. of laterals × No. of emitters/lateral

$$= 168 \times 1788$$

$$= 300384$$

Total wetted area by emitters = 0.22 × 300384

$$= 66084.48 \text{ m}^2$$

$$= 6.61 \text{ ha}$$

Total area of field = 21.45 ha

$$\text{Percentage wetted area} = \frac{6.61}{21.45} \times 100 = 30.81\%$$

Considering a flat percentage wetted area as 35%,

Net depth of water application

$$= (FC - WP) \times AS \times Zr \times 10 \times MAD \times WA$$

$$= (16 - 6) \times 1.4 \times 0.80 \times 10 \times 0.20 \times 0.35$$

$$= 7.84 \text{ mm}$$

If the efficiency of the system is 95%, $E_s = 0.95$

$$\text{Gross depth of water application} = \frac{7.84}{0.95} = 8.25 \text{ mm}$$

Maximum crop evapotranspiration is already known as 6.72 mm/day.

$$\text{Irrigation interval} = \frac{8.25}{6.72} = 1.23 \text{ days} \approx 30 \text{ hrs}$$

$$\text{Application rate of the system} = \frac{q}{S_e \times S_l}$$

$$= \frac{4}{0.4 \times 1.8}$$

$$= 5.55 \text{ mm/hr}$$

$$\text{Duration of water application} = \frac{\text{Gross depth of water application}}{\text{Application rate}}$$

$$= 8.25/5.55$$

$$= 1.48 \text{ hr}$$

For deciding the number of sub-plots, consider 80% of irrigation interval, it comes 24 hrs.

No. of sub-plots

$$= \frac{80\% \text{ of irrigation interval}}{\text{Duration of water application}}$$

$$= \frac{24}{1.48} = 16.21 \approx 16 \text{ sub-plots}$$

Design of Lateral Pipe:

Choose lateral pipe of 20 mm diameter.

Working pressure = 30 m

Length of one lateral pipe in one sub-plot $= \dfrac{715}{8} = 89.37 \approx 89.5$ m

No. of drippers on each lateral $= \dfrac{89.50}{0.4} = 224$

Maximum allowable head loss in the mainline = 20% of working pressure

$$= 30 \times \frac{20}{100} = 6 \text{ m}$$

Discharge flowing through each lateral Q = 4 lph × 224

$$= 0.004 \text{ m}^3/\text{h} \times 224 = 0.90 \text{ m}^3/\text{h}$$

$$= 0.25 \text{ lps}$$

Head loss due to friction along the 89.5 m of the lateral line,

$$h_f = K \frac{(Q/C)^{1.852}}{D^{4.87}} \times \frac{L}{100}$$

$$= 1.22 \ x \ 10^{12} \times \frac{(0.25/130)^{1.85}}{(20)^{4.87}} \times \frac{89.5}{100}$$

$$= 4.76 \text{ m}$$

If we consider local heal loss as 10% of loss along the lateral (h_f)

$$h_f = h_f \times 1.10 \times F$$

$$= 4.76 \times 1.10 \times 0.366$$

$$= 1.92 \text{ m}$$

This head loss of lesser than 6 m, so 20 mm will be selected.

Therefore, total pressure required at the inlet of the lateral

$$= hs + 0.75 \times h_f$$

$$= 30 + 0.75 \times 1.92$$

$$h_{f\text{emitter}} = 31.43 \text{ m}$$

Design of the Manifold:

Only one side of the manifold is connected with laterals.

No. of lateral pipes in manifold towards slope $= \dfrac{300/2}{1.8} = 83.33 \approx 84$

Mainline is laid out in the middle of the field.

Length of manifold $= \dfrac{300}{2} = 150$ m

Elevation difference $= 0.50$ m

Discharge carried out by one manifold

Q = No. of laterals × Discharge passing through one lateral

$= 84 \times 0.90 = 75.6$ m^3/h

Select the size of manifold as 140 mm PVC pipe (C =140)

Head loss due to fixation

$hf = K\dfrac{(Qlc)^{1.852}}{D^{4.87}} \times \dfrac{L}{100} \times F$ $= 1.22 \; x \; 10^{12} \times \dfrac{(21/140)^{1.85}}{(140)^{4.87}} \times \dfrac{150}{100} \times 0.366$ $= 0.708$ m	$Q = 75.6 \text{ m}^3/\text{h}$ $= \dfrac{75.6 \times 1000}{3600} = 21 \text{ lps}$

Considering 10% of head loss along the sub-main

Total head loss along the manifold = 0.708 × 1.10 = 0.77 m

Now pressure required at the entrance of the manifold , H_{manifold}

$= h_{fe} + 0.75 \; hf - (Z/2)$

$= 31.43 + 0.75 \times 0.77 - (0.5/2)$

$H_{\text{manifold}} = 31.43 + 0.58 - 0.25 = 31.76$ m

Design of main line:

Total number of manifold $= 16$

Number of manifold working at a time $= 1$

Length of mainline $= 7 \times 89.85 = 626.5$ m

Here we can go for telescopic design of mainline.

First 400 m length is of 225 mm dia and remaining 226.5 m is of 140 mm dia.

For 400 m length with 225 mm pipe dia, discharge carried out by the pipe is = 75.6 m^3/h

(Since one manifold is operating at a time, mainline will carry only discharge of one manifold anytime.)

Pressure head loss in mainline,

$$h_f = 1.22 \times 10^{12} \times \frac{(21/140)^{1.85}}{(225)^{4.87}} \times \frac{400}{100} \times 0.366 = 0.18 \text{ m}$$

For remaining 226.5 m length of pipe with 140 mm dia

$$h_f = 1.22 \times 10^{12} \times \frac{(21/140)^{1.85}}{(140)^{4.87}} \times \frac{226.5}{100} \times 0.366 = 1.07 \text{ m}$$

Total head loss along the mainline = 0.18 + 1.07 = **1.25 m**

Considering 10% loss in the mainline,

Total frictional head loss, h_f = 1.25 × 1.10 = **1.38 m**

Again considering the elevation difference as 1 m,

Total head loss at the inlet of mainline

$$= H_{\text{manifold}} + 0.75 \times 1.38 + ½$$
$$= 31.76 + 1.04 + 0.5$$
$$= 33.32 \text{ m}$$

Pump Selection:

Considering the head loss in control head is 10% of total head loss

$$= 33.3 \times 1.10 = \textbf{36.63 m}$$

Required discharge capacity of the pump = 89.5 m³/h

Required horse power
$$= \frac{Q \times H}{273 \times \text{Efficiency of pump}}$$
$$= \frac{89.5 \times 36.63}{273 \times 0.75}$$
$$= \textbf{16 hp}$$

Example 7.9. Design a drip irrigation system for a 2 ha Dwarf Cavendish banana plantation in Ranchi. The field is assumed to be rectangular with a length of 200 m and a width of 100 m. The field is nearly flat and the soil is sandy loam. The source of water is located at left bottom corner of the field. The plant spacing is 2 × 2 m and the maximum daily pan evaporation is 9 mm. The crop coefficient of banana at middle stage is 1.2.

Sol.:

Layout of drip irrigation system

The mainline is made along the length of the field up to 100 m. The submain was laid out in the middle of the field and both sides, laterals are connected directly to the sub-main. The laterals are spaced 2 m apart on each row of the plants.

Estimation of water requirement

The maximum daily pan evaporation is 9 mm. The average canopy coverage of the plant is assumed to be 60%.

Volume of water to be applied **daily** = Crop coefficient × Pan coefficient × Percentage of wetted area × Daily pan evaporation × Area to be irrigated

$$= 1.2 \times 0.70 \times 0.60 \times {}^9/_{1000} \text{ m} \times (200 \times 100 \text{ m}) = 90.72 \text{ m}^3 = 90720 \text{ litres}$$

Capacity of drip irrigation system,

Assuming water application efficiency as 95% and daily irrigation water supply for 3 hrs in a day. The capacity can be determined as

$$\frac{\text{Volume of water to be applied daily}}{\text{Duration of irrigation} \times \text{Application efficiency}} = \frac{90720 \times 1}{3 \times 0.95} = 31832 \text{ lph} = 8.84 \text{ lps}$$

Number of lateral and drippers

Number of laterals $= \dfrac{100}{2} = 100$ and the length of each lateral is 100 m

Number of banana plants per lateral $= \dfrac{100}{2} = 50$

Total number of plants to be irrigated at a time = 50 × 100 = 5000

Discharge required per plant per day $= \dfrac{31832}{5000} = 6.37 \text{ lph}$

Drippers with a discharge rate of 1, 2 and 4 lph for one plant may be selected. Drippers can be placed in triangular shape with the help of microtube.

Selection of Diameter of Lateral Pipe

Let us take 16 mm lateral pipe and length of each lateral is 100. We need to determine the friction loss and it should be less than the 10% of operating pressure head for dripper. Therefore, any friction loss less than 1m will work. Hazen-Williams formula will be used to determine the friction loss and is given hereunder:

$$h_f = 1.212 \times 10^{12} \frac{\left(Q/C\right)^{1.852}}{D^{4.87}} \frac{L}{100}$$

where

h_f = frictional head loss in the pipeline (m)

Q = discharge in pipeline (lps)

C = friction coefficient, is the length of pipeline

D = inside diameter of the pipeline (mm)

In this example, we have Q = 50 × 7 = 350 lph = 0.10 lps, C = 130,

$$L = 100 \text{ m}, D = 16 \text{ mm}$$

How, we can calculate the friction head loss in the lateral pipe:

$$h_f = 1.212 \times 10^{12} \frac{\left(0.10\big/130 \right)^{1.852}}{16^{4.87}} \frac{100}{100} = 2.83 \text{ m}$$

For D = 12 mm, h_f = 11.50 m

For D = 20 mm, h_f = 0.95 m

As 20 mm diameter pipeline yields friction loss less than 1 m, we will choose lateral diameter as 20 mm.

Selection of Diameter of Sub-mainline

Sub-main will carry all the discharge to be carried out by all the laterals. There are 100 laterals on both sides of sub-mainline and each lateral is carrying a discharge of 0.10 lps. We will apply the same Hazen–Williams equation:

$$h_f = 1.212 \times 10^{12} \frac{\left(Q\big/C \right)^{1.852}}{D^{4.87}} \frac{L}{100}$$

Here Q = 10 lps, C = 150, L = 100 m and let us assume D = 75 mm

$$h_f = 1.212 \times 10^{12} \frac{\left(10\big/150 \right)^{1.852}}{75^{4.87}} \frac{100}{100} = 5.94 \text{ m}$$

For, D = 100 mm, h_f = 1.46 m

D = 125 mm, h_f = 0.49 m

Christiansen (1942) developed a reduction coefficient '*F*' which will take care of reducing discharge along the sub-mainline. In this equation: *m* is flow exponent in the head loss equation, i.e. 1.852. *N* is the number of outlets. In this case, the sub-main is having 100 outlets through which laterals extend. Therefore,

$$F = \frac{1}{1.852 + 1} + \frac{1}{2 \times 100} + \frac{(1.852 - 1)^{0.5}}{6 \times 100^2} = 0.35$$

This needs to be multiplied by calculated for sub-mainline. We get

For, D = 100 mm, Final h_f = 1.46 × 0.35 = 0.51 m

D = 125 mm, Final h_f = 0.49 × 0.35 = 0.17 m

D = 75 mm, Final h_f = 5.94 × 0.35 = 2.07 m

'Though, the friction loss is lowest in the case of D = 125 mm but it is well within the limit for D= 100 mm. We will choose D = 100 mm in order to save the cost.'

Similarly, the main pipeline of length 100 m will carry the same discharge as sub-main pipeline is carrying, i.e. 10 lps.

Let us assume, D = 125 mm

$$h_f = 1.212 \times 10^{12} \frac{\left(10 / 150\right)^{1.852}}{125^{4.87}} \frac{100}{100} = 0.49 \text{ m}$$

We will choose 125 mm diameter pipeline for mainline.

Head at the inlet of the sub-main = $H_{emitter} + H_{lateral} + H_{(sub-main)} + H_{slope}$
$$= 10 + 0.95 + 0.51 + 0 = 11.46 \text{ m}$$

Pressure head variation along sub-main = $\dfrac{11.46 - 10.95}{11.46} = 0.044$, i.e. 4.4 %.
That is within the limit of 20%.

Determination of size of the pump

Total dynamic head, $H = H_{emitter} + H_{lateral} + H_{(sub-main)} + H_{main} + H_{slope}$
$$+ H_{static} + H_{local}$$
$$= 11.46 + 0.49 + 10 + 1.19 = 23.14 \text{ m}$$

H_{local} is 10% of $H_{emitter} + H_{lateral} + H_{sub-main} + H_{main} + H_{slope}$

If efficiency of the pump is 60%, then

Pump horse power, $\text{hp} = \dfrac{H \times Q}{75 \times \eta_p} = \dfrac{23.14 \times 10}{75 \times 0.60} = 5.14 \text{ hp}$

A drip irrigation system of the above layout and specification of laterals, sub-main and main pipeline is acceptable. Cost can be manipulated by changing the layout of the system.

Example 7.10. Determine the cost of energy to pump water for drip irrigated banana plantation for 25 ha of land with an irrigation system that requires a 30 m of pressure head. Assume that the pump operates at 85% efficiency and drip system water application efficiency is 90%. The cost of energy is ₹2.50/kW-hr. Assume that banana plantation requires 2 m of water per year.

Sol.:

Total discharge to delivered to the field = $\dfrac{2 \times 25 \times 10000 \times 1000}{365 \times 24 \times 3600} = 15.85 \text{ lps}$

Horse power $= \dfrac{Q \times H}{75 \times \text{Application efficiency} \times \text{Pump efficiency}}$
$$= \dfrac{15.85 \times 30}{75 \times 0.90 \times 0.85} = 7.84 \text{ hp} = 5.84 \text{ kW}$$

Cost of energy to pump water for one hour = 2.50 × 5.84 = ₹14.60

Ans. ₹14.60

7.6. Design Tips for Drip Irrigation

Though the design methodology has been explained previously, there are some points which should be considered while planning a drip irrigation system for crop production. As drip system is also used for fertigation and chemical injection, application uniformity of fertilizers and chemicals will be affected with water application efficiency of the drip irrigation systems. Here are some points.

- Length of lateral lines should not exceed the recommended length. Excessive length of laterals will result in poor uniformity and uneven water application. For better precaution, check the amount of water supplied by the emitters in the last section of the lateral.

- The sizes of main and sub-mainlines should be properly selected to avoid excessive pressure losses and velocities. Excessive velocities of water in the lines, due to too small a diameter, can create a water hammer which can damage the main and sub-mainlines.

- Use pressure compensating emitters where necessary.

- Choose the irrigation area zone in conformity with the existing water pump.

- The drip irrigation systems experience clogging of emitters which reduce water application uniformity. The degree of clogging depends on water quality, type of emitters, and quality of water filtration. An irrigation system should include an injection port to allow for injection of chlorine for flushing out the systems at regular interval.

- The irrigation system injecting fertilizers or chemicals is required to be provided with proper backflow mechanism to prevent the contamination of the water source.

- Pressure changes in the irrigation line should be monitored. An increase in pressure can be caused by emitter clogging or some other blockage such as the failure of a valve or pressure regulator. A pressure drop may indicate a broken line, leaking valves, the failure of flush valves to close properly.

7.7. Economic Analysis

Investment on installation of drip irrigation system must be economically viable and justified. If the project gives economic surplus, we can say the project is economically viable. As drip irrigation involves considerable cost, the economic appraisal of the installation of drip system must take full account of all the cost and benefit likely to be accrued from the crops to be grown and its byproduct. Economic analysis is carried out to determine whether the returns from the project will be able to justify the investment or not. Drip irrigation project cost will include all the expenditure made on procurement, installation, operation and cultivation cost. The annual cost of a project includes both fixed and variable costs. However, at the policy level, environmental contamination and sustainability should be a part of the analysis of irrigation systems and irrigation projects.

The benefits of irrigation through drip irrigation are many such as better crop survival, earlier fruit production, more yields, efficient distribution of nutrients, less plant stress, reduced yield variability, and improved fruit quality. We will present herein the methodology applied to evaluate the economics of irrigation. Growers/farmers operating drip irrigation must identify the drip irrigation investment, operating cost or yield response. Before taking up analysis part, it is essential to understand the few basic terms and its determination technique.

The purpose of cost-benefit analysis is to assess the economic merit of any project which requires funding. These long-term projects used to run for several years, so they will have cost and benefit flows every year. For this reason, cost and benefits analysis cannot be done by simply adding or subtracting the cost and benefits. When we invest a sum in a 'project', this is called the principal amount and this principal amount will earn interest as we move on. The interest accrued so far will also attract interest. This is compounding of interest. Compound interest formula is given by

$$F = P(1 + i)^t$$

where,

F = value of a sum in some future years (n)

P = value of a sum in the present year

t = number of years

i = interest rate

We normally say for convenience that rupees outflow or inflow occur at the end of a period. Now suppose, ₹1,00,000 is invested for 5 years with rate of interest as 8%. After 5 years, you will get

$$F = 100000 (1 + 0.08)^5 = 1,46,932.80$$

You can compound interest yearly, quarterly or monthly. The more frequently you compound, the more interest you earn, but often the increase is not much. Suppose, we compound the interest quarterly. We need to multiply the number of years by 4 and divide the interest rate by 4.

$$F = 100000(1 + 0.02)^{20} = 1,48,594.70$$

There is an increase in return.

7.7.1 Payback Period

Cash inflow is the amount which is going into a business like sales of produce, investments or financing. It is the opposite of cash outflow, i.e. the money is leaving the business. A project is considered good, if the cash inflow is greater than the cash outflow. Net cash flow is the cash inflows minus cash outflows over a given period and it may be positive or negative. The cumulative cash flow is determined by adding all of the cash flows from the inception of the project. Analysis of the cumulative cash flow can help us in understanding the long-term strength of the project.

The payback period is the time required to recover the initial investment. It is the number of years in which the initial investment made for a project is recovered. The project with the least number of years usually is selected. It ignores the time value of money. Time value of money means that ₹100 today is more valuable than ₹100 tomorrow. To overcome this, discounted cash flow technique is preferred which we have discussed. The formula for computing payback period with even cash flows is:

$$\text{Payback period} = \frac{\text{Total outflows}}{\text{Inflow every year}} \quad \text{or} \quad \frac{\text{Initial investment}}{\text{Net annual cash inflows}}$$

Let us take two projects 'X' and 'Y'. The initial investment in both the projects is ₹5,00,000. Project X has even inflow of ₹50,000 every year and project Y has uneven cash flows as follows:

Year 1—₹1,00,000, Year 2—₹1,50,000, Year 3—₹2,00,000,

Year 4—₹50,000

$$\text{Payback period for project X} = \frac{\text{Initial investment}}{\text{Net annual cash inflows}} = \frac{500000}{50000} = 10 \text{ years}$$

In case of project Y, the total inflows (1,00,000 + 1,50,000 + 2,00,000 + 50,000) becomes equal to total outflows in 4 years. Therefore, the project Y takes 4 years to recover the initial investment. However, the project 'X' takes 10 years to recover the investment, certainly the project 'Y' is preferred.

7.7.2 Depreciation

It is a well known fact that if you buy a product today, its value after 3–4 years will not be the same. This means that the product has been depreciated due to use over the years. Depreciation is the value reduction of any asset due to physical use over the time periods. The annual depreciation is calculated from the following formula.

$$\text{Annual depreciation} = \frac{\text{Primary cost} - \text{Salvage value}}{\text{Useful life in years}}$$

Salvage value is an estimate of the remaining value of an investment at the end of its useful life. It is desirable to determine the present worth of a future value of a product, which is called discounting, for any analysis of a project which is going to be operative for a ling-time. The present worth of a future value of a product at the end of n years at an interest rate of i can be computed by using the following expression.

$$PW = F\left(\frac{1}{(1+i)^n}\right) \qquad \qquad ...(7.44)$$

where,

PW = present worth of the future income value

F = future values of the income

7.7.3 Escalation Cost

The rate of escalation can be incorporated in the analysis of present worth and annual cost. If e is the annual rate of escalation, the present worth value, which incorporates the effect of escalation in the cost, can be estimated by the following formula.

$$PW(e) = PW \times \left(\frac{(1+e)^n}{(1+i)^n}\right) \qquad \qquad ...(7.45)$$

Keller and Bliesner (1990) considered the interest rates (i), the expected life of investment 'n' and an estimate of the expected annual rate of escalation in the calculation of annual energy cost. The present worth of the escalating energy factor and the equivalent annualized cost of escalating energy factor were computed by the following equations:

$$PW(e) = \left[\frac{(1+e)^n - (1+i)^n}{(1+e) - (1+i)}\right]\left[\frac{1}{(1+i)^n}\right] \qquad ...(7.46)$$

and equivalent annualized cost of escalating energy (EAE) factor at annual rate i is calculated as:

$$EAE(e) = \left[\frac{(1+e)^n - (1+i)^n}{(1+e) - (1+i)}\right]\left[\frac{i}{(1+i)^n - 1}\right] \qquad ...(7.47)$$

where,

$PW(e)$ = present worth factor of escalating energy cost taking into account the time values of money over the life cycle.

$EAE(e)$ = equivalent annualized cost factor of escalating energy taking into account the value of money over life cycle.

7.7.4 Capital Recovery Factor

A capital recovery factor (CRF) converts a present value into a stream of equal annual payments over a specified time at a given discount rate (interest rate). If P is the present value of a product, then amount of each level payment to be made at the end of each of n periods can be determined by multiplying it with *CRF*. The standard capital recovery factor CRF is computed by

$$CRF = \frac{i(1+i)^n}{(1+i)} \qquad \qquad ...(7.48)$$

7.7.5 Discounting Methodology

Sometimes, the compounding may be a case of continuous compounding. In this case, the future value, F, can be calculated by:

$$F = P.e^{i.t} \qquad ...(7.49)$$

Let us calculate with same data given above,

$$F = 100000 . e^{0.08 \times 5} = 1,49,182.40$$

In other words, if we get ₹149182.40 after 5 years, its present value is ₹100000.

When we talk about discounting, it is the reverse of compounding. We diminish the future value to a present value by discounting. Discounting is an important concept for benefit cost analysis of any investment project. The interest rate will be termed discount rate. The discounting formula is written as:

$$PW = F\left(\frac{1}{(1+i)^n}\right) \qquad ...(7.50)$$

where, PW is present worth of the future income value. Discounting with a positive rate of discount always will diminish the initial amount. If the initially investment (P or PW) and investment accumulated after n years (F), the annual rate of earning, i.e. i can be determined. Time and discount rates can have a major impact on present value calculations.

While analyzing the drip irrigation project, when we convert all the cost and benefits to a common time base, we call it discounted cash flow technique. In short, all the costs and benefits are compared on the basis of a common time scale, though this may occur at different time periods. Every investment project will have cost and benefit. Based on simple knowledge of total cost and total benefit, we can measure three indicators of economic evaluation such as net present value (NPV), benefit-cost ratio (BCR), and internal rate of return (IRR). We will make it clear with a solved example in this chapter.

7.7.5.1 Net Present Value (NPV)

This is a single value, representing the difference between the sum of the projected discounted cash inflows and outflows attributable to a capital investment, using a discount rate that properly reflects the relevant risks of those cash flows. Using NPV as indicator, we convert all the cost and benefit of any year into present year. If the value of NPV is positive, the project benefit has more than the cost, and then the project is feasible and can be taken up for implementation. It may be interpreted as the present worth of the income generated by the investment. The NPV can be calculated as follows:

$$NPV = \Sigma_{t=0}^{n} \frac{B_t - C_t}{(1=i)^t} \qquad ...(7.51)$$

where,

B_t = benefit at time t,

C_t = cost at time t,

i = discount rate,

n = number of years.

7.7.5.2 Benefit Cost Ratio (BCR)

The discounted measure of the project worth can be expressed by benefit-cost ratio. This is the ratio that implies the return per rupee investment. The benefits and costs of any year are converted into the equivalent basis (i.e. present year) to find out the benefit–cost ratio. If BCR value is greater than 1, benefit has more than the cost, then the project is feasible. The B/C ratio can be worked out by using the following formula:

$$BCR = \frac{\sum_{t=0}^{n} \frac{B_t}{(1+i)^t}}{\sum_{t=0}^{n} \frac{C_t}{(1+i)^t}} \qquad ...(7.52)$$

7.7.5.3 Internal Rate of Return (IRR)

The average annual percentage return is expected from a project, where the sum of the discounted cash inflows over the life of the project is equal to the sum of the discounted cash outflows. Therefore, the IRR represents the discount rate that results in a zero NPV of cash flows. In this method also, we convert all the costs as well as benefits of any year into equivalent basis (i.e. present year). IRR can be compared with the existing bank interest rate to judge the economic feasibility of the project. IRR is calculated from the following principle.

$$\sum_{t=0}^{n} \frac{B_t}{(1+i)^t} = \sum_{t=0}^{n} \frac{C_t}{(1+i)^t} \qquad ......(7.53)$$

The cost of the drip irrigation system includes all of the fixed costs, operation costs, maintenance costs and all the costs incurred to the project. A point where profile of net present value crosses horizontal axis at plotted graph indicates project internal rate of return. The benefit of the investment will include, income from production, any form of by-product, salvage value, etc.

A major and limiting factor is fixed cost which determines the economic feasibility of the drip irrigation system for any crop. The close spacing crops need higher fixed investment and wide-spaced crops require relatively lower capital investment. Most states including Tamil Nadu are providing about 50% of the capital cost as subsidy either through a state sponsored scheme or centrally sponsored scheme to encourage the adoption of drip irrigation for different crops since it is a capital-intensive technology. The average capital cost of drip-set for okra crop comes to about ₹38,533/acre without subsidy, whereas it is only ₹27,993/acre after deducting the state subsidy (Narayanamoorthy and Devika, 2017).

Example 7.11. A drip irrigation project is being planned with a rough estimate. The tentative income and expenditure of the farmer for 5 years is given in the table.

Year	Cash inflow	Cash outflow
1	0	5,00,000
2	3,00,000	90,000
3	4,00,000	1,00,000
4	1,00,000	1,75,000
5	50,000	35,000

What is the payback period for this project and what is the net cash flow at the end of 5 years? If the net present value for each of the cash flows are calculated at a 10% interest rate, calculate the net present value cash flow at the end of 5 years. Compare the net present value cash flow with total cash flow without the net present value applied.

Sol.: The following table shows the calculations of the net cash flow and cumulative cash flow (₹).

Year	Cash inflow	Cash outflow	Net cash flow	Cumulative cash flow
1	0	5,00,000	5,00,000	5,00,000
2	3,00,000	90,000	2,10,000	2,90,000
3	4,00,000	1,00,000	3,00,000	10,000
4	1,00,000	1,75,000	75,000	65,000
5	50,000	35,000	15,000	50,000

If we see the above Table, it is clear that in third year, the cash inflow is exceeding the cash outflow, hence the payback period of the project is 3 years. The net cash flow at the end of 5 years is ₹50,000/- (row 5 of cumulative cash flow). Let us calculate the net present value for each of the cash flows at a 10% (0.01) interest rate

$$NPV = \left(-\frac{500000}{(1+0.1)^0} \right) + \left(-\frac{290000}{(1+0.1)^1} \right) + \left(\frac{10000}{(1+0.1)^2} \right)$$

$$+ \left(-\frac{-65000}{(1+0.1)^3} \right) + \left(-\frac{50000}{(1+0.1)^4} \right)$$

$$= -\,838{,}358$$

We have the cash outflow without NPV = $-500{,}000 - 290{,}000 + 10{,}000$

$$- 65{,}000 - 50{,}000 = -895{,}000.$$

Therefore, the net present value cash flow at the end of 5 years is less than the total cash flow without the net present value applied

Example 7.12. A grape grower is planning to adopt the drip irrigation system for 15 ha of land. An initial investment of ₹10,00,000 was speculated. It was determined to make incomes of ₹2,50,000 in the primary year after the end of the project and of ₹3,20,000 in each of the next two years. Explain whether the project is attractive over the 3 years cycle at an interest rate of 10%.

Sol.: The project is expected to generate revenues of ₹2,50,000 in the first year after the end of the project and of ₹3,20,000/- in each of the two following years. We need to determine the present value (PW) for 3 years. Using the same formula for each year,

$$PW = F\left(\frac{1}{(1+i)^n}\right) = 250000\frac{1}{(1+0.10)^0} = 250000.00$$

For $n = 1$

$$PW = 320000\left(\frac{1}{(1+0.10)^1}\right) = 290909.00$$

For $n = 1$

$$PW = 320000\left(\frac{1}{(1+0.10)^2}\right) = 264462.00$$

The total present worth becomes ₹250000 + 290909 + 264462 = ₹805371. If we subtract the initial investment (10,00,000) from total present worth, it becomes negative. As net present value is negative, the project is unattractive.

Example 7.13. Tomato crop is to be cultivated in 1 ha area under drip irrigation system. Determine the net present value (NPV) and benefit-cost ratio (BCR). The cash flow pattern is given hereunder and take interest rate as 12% and life of the drip system as 10 years. The operation cost include the maintenance cost of drip systems and cost of cultivation which may increase over the period of time, however the return, i.e. cash inflow has been assumed constant with a yield of tomato as 30 t/ha.

Year	Fixed cost	Operation and maintenance	Cash inflow	Year	Fixed cost	Operation and maintenance	Cash inflow
1	180000	40000	120000	6	—	50000	120000
2	—	40000	120000	7	—	50000	120000
3	—	40000	120000	8	—	60000	120000
4	—	45000	120000	9	—	60000	120000
5	—	45000	120000	10	—	60000	120000

Sol.: First we will calculate the discount rate which is also called the discount factor with the following formula

$$\text{Discount factor} = \left(\frac{1}{(1+i)^n}\right)$$

here, n is 1, 2, 3, ..., 10.

Cash outflow   = Fixed cost + Operation and maintenance

Cash flow   = Cash inflow – Cash outflow

Discounted cash flow   = Cash flow × Discount factor

Discounted cash outflow = Cash outflow × Discount factor

Discounted cash inflow   = Cash inflow × Discount factor

The result is presented in a tabular form.

Year	Fixed cost	Operation and maintenance	Cash outflow	Cash inflow	Cash flow	Discount factor	Discounted cash flow	Discounted cash outflow	Discounted cash inflow
1	180000	40000	220000	120000	–100000	0.8929	–89290	196438	107148
2		40000	40000	120000	80000	0.7972	63776	31888	95664
3		40000	40000	120000	80000	0.7118	56944	28472	85416
4		45000	45000	120000	75000	0.6355	47662	28597	76260
5		45000	45000	120000	75000	0.5674	42555	25533	68088
6		50000	50000	120000	70000	0.5066	35462	25330	60792
7		50000	50000	120000	70000	0.4523	31661	22615	54276
8		60000	60000	120000	60000	0.4039	24234	24234	48468
9		60000	60000	120000	60000	0.3606	21636	21636	43272
10		60000	60000	120000	60000	0.322	19320	19320	38640
Total			670000	1200000			253960	424063	678024

Net present value (NPV) = Sum of discounted cash flow for the

periods of 10 years

= ₹2,53,960

Since NPV is greater than zero, investment on drip irrigation is economically feasible. Now, calculate benefit cost ratio (BCR).

$$\text{Discounted benefit cost ratio} = \frac{\text{Discounted cash inflow}}{\text{Discounted cash outflow}}$$

$$= \frac{678024}{424063}$$

$$= 1.60$$

Now, what will happen to BCR, if discounted cash flow technique is not followed in economic analysis?

$$\text{Undiscounted cash flow technique} = \frac{\text{Cash inflow}}{\text{Cash outflow}}$$

$$= \frac{1200000}{670000}$$

$$= 1.80$$

BCR is higher in case of undiscounted cash flow technique which may not be appropriate techniques of project appraisal and feasibility study.

Example 7.14. Determine the internal rate of return of the investment made in covering 1 ha mango (Amrapali) orchard under drip irrigation. The cash flow patterns are given hereunder and take interest rate as 12% and life of the drip system as 10 years. The operation cost include the maintenance cost of drip systems and cost of cultivation which may increase over the period of time, however, the return, i.e. cash inflow has been assumed constant.

Year	Fixed cost	Operation and maintenance	Cash inflow	Year	Fixed cost	Operation and maintenance	Cash inflow
1	75000	10000	—	6	—	12000	50000
2	—	10000	—	7	—	12000	50000
3	—	10000	—	8	—	15000	50000
4	—	12000	50000	9	—	15000	50000
5	—	12000	50000	10	—	15000	50000

Sol.: As in the previous example, we will calculate the discount rate which is also called the discount factor with the following formula for 12, 14, 18% of interest rate.

$$\text{Discount factor} = \left(\frac{1}{(1+i)^n}\right) \text{ here, } n \text{ is } 1, 2, 3, ..., 10.$$

Cash inflow has not been considered for first 3 years and from fourth year onwards, we have included cash inflows. Again, it has been assumed that cash inflows throughout the life cycle is constant, of course it is not true. The result is presented in a tabular form.

Year	Fixed cost	Operation and maintenance	Cash outflow	Cash inflow	Cash flow	Discounted cash flow		
						12%	14%	18%
1	75000	10000	85000	0	−85000	−75896.5	−74561.4	−72033.9
2		10000	10000	0	−10000	−7972	−7694.68	−7181.84
3		10000	10000	0	−10000	−7118	−6749.72	−6086.31
4		12000	12000	50000	38000	24149	22499.05	19599.98
5		12000	12000	50000	38000	21561.2	19736.01	16610.15
6		12000	12000	50000	38000	19250.8	17312.29	14076.4
7		12000	12000	50000	38000	17187.4	15186.22	11929.15
8		15000	15000	50000	35000	14136.5	12269.57	9311.336
9		15000	15000	50000	35000	12621	10762.78	7890.962
10		15000	15000	50000	35000	11270	9441.033	6687.256
Total			198000	350000	198000	**29189.4**	**18201.1**	**803.1**

As NPV of ₹29189.4 at 12% interest rate is highest and greater than zero, the investment is economically good.

IRR = Lower discount rate + Difference between the two discount rate ×

$$\left(\frac{\text{NPV lower rate}}{\text{Abs. diff. between NPV at two discount rates}}\right)$$

$$= 14 + 4 \times \left(\frac{18201}{17398}\right)$$

$$= 18.18 = \mathbf{18\%}$$

If discount rate of 12% and 14% is considered. Irr is obtained as 17.3%

This is the minimum discount rate that we use to identify what our investment will yield an acceptable return.

■■

8

PERFORMANCE EVALUATION OF EMISSION DEVICES

The performance of emission devices has a vital role for the success of a drip irrigation system. It is therefore necessary to test the emitters so that the designers can select the best suitable emitters for a specific requirement and design the system accordingly. The different characteristics used for performance evaluation of drip emitters are described below.

8.1 Manufacturing Characteristics

The manufacturing characteristics of emitters are described by manufacturing coefficient of variation and mean flow rate variation.

8.1.1 Manufacturing Coefficient of Variation

This is a parameter which is used to measure the emitter flow variation caused by manufacturing defects. Common causes of this variation are the inability to hold dimensional tolerances due to molding pressures and temperature and variation in the materials used. This statistical parameter was given by Keller and Karmeli (1974) and expressed as

$$C_v = \frac{s_d}{q_{avr}} \qquad \qquad ...(8.1)$$

where,

C_v = manufacturing coefficient of variation

s_d = standard deviation of the sample

q_{avr} = average emitter flow rate, lph

The American Society of Agricultural Engineers (ASAE) and Indian Standard Institution (IS:10799) have provided interpretation of classification of emitters based on the manufacturing coefficient of variation (Table 8.1).

Table 8.1: Recommended classification of manufacturing coefficient of variation

Emitter type	*C_v range*	*Classification*
Point source	<0.05	Good
	0.05 to 0.10	Average
	0.10 to 0.15	Marginal
	>0.15	Unacceptable
Line source	<0.10	Good
	0.10 to 0.20	Average
	>0.20	Marginal to unacceptable

8.1.2 Mean Flow Rate Variation

Besides the individual emitter flow rate variation, the measured mean flow rate varies from nominal flow rate. The percentage difference between actual flow rate and nominal flow rate is characterized by mean flow rate variation and can be expressed as.

$$Q_d = \frac{(q_r - q_{avr})}{q_r} \times 100 \qquad \qquad ...(8.2)$$

where,

Q_d = mean flow rate variation (%)

q_r = nominal emitter flow rate (lph)

q_{avr} = average emitter flow rate (lph)

8.2 Hydraulic Characteristics

Keller and Karmeli (1974) have shown that the emitter flow can be characterized by

$$q = K_d H^x \qquad \qquad ...(8.3)$$

where,

q = emitter flow rate (lph)

K_d = coefficient of discharge

H = pressure head at the emitter (kg/cm^2)

x = emitter discharge exponent

The value of x characterizes the flow regime and relationship between discharge versus pressure of the emitter. The lower the value of x, the less discharge will be affected by the pressure variation. The flow in simple orifice and nozzle emitters is turbulent where $x = 0.5$. For fully compensating emitters, $x = 0.0$. The exponent for long path emitter varies from 0.7 to 0.8. For vortex emitters,

x is about 0.4. The exponent x for tortuous path emitters usually falls between 0.5 and 0.7. The relationship between the emitter flow variation and the pressure variation is given by

$$q_{var} = 1 - (1 - H_{var})^x \qquad \qquad ...(8.4)$$

and

$$H_{var} = \frac{H_{max} - H_{min}}{H_{max}} \qquad \qquad ...(8.5)$$

where,

$$q_{var} = \text{emitter flow variation}$$
$$H_{var} = \text{pressure variation}$$
$$H_{max} \text{ and } H_{min} = \text{maximum and minimum pressure in the line}$$

When the value of x is 0.5, true for most of the orifice type of emitters, a pressure variation of 20% is equivalent to a 10% emitter flow variation, and a pressure variation of 10% is equivalent to a 5% emitter flow variation. The emission uniformity is a quantitative expression of the emitter flow variation. Another way of calculating the emitter flow variation is by comparing the maximum with minimum emitter flow. The emitter flow variation is calculated by,

$$q_{var} = \frac{q_{max} - q_{min}}{q_{max}} \qquad \qquad ...(8.6)$$

where, q_{max} and q_{min} = maximum and minimum emitter flow variation along the line, respectively.

8.3 Operational Characteristics

The operational characteristic of emitters includes the uniformity of water emission from emitters. Various formulae for estimating emission uniformity are given below.

8.3.1 Emission Uniformity

Emission uniformity is a measure of the uniformity of emissions from all the emission points within an entire drip irrigation system. As per Keller and Karmeli (1974), the emission uniformity is given by

$$EU = \frac{q_n}{q_{avr}} \times 100 \qquad \qquad ...(8.7)$$

where,

EU = emission uniformity

q_n = average of lowest one–fourth of emitter flow rate (lph)

q_{avr} = average emitter flow rate (lph)

8.3.2 Absolute Emission Uniformity

Keller and Karmeli (1974) proposed the following formula for estimating absolute emission uniformity given hereunder.

$$EU_a = 1/2 \left(\frac{q_n}{q_{avr}} + \frac{q_{avr}}{q_x} \right) \times 100 \qquad ...(8.8)$$

where,

q_n = average of lowest one-fourth of emitter flow rate (lph)

q_{avr} = average emitter flow rate (lph)

q_x = average of highest one-eighth of emitter flow rate (lph)

8.3.3 Emission Uniformity

Karmeli and Keller (1975) modified the above proposed emission uniformity formula for testing the drip irrigation system in the field as

$$EU = 100 \times \left[1 - 1.27 \frac{C_v}{\sqrt{n}} \right] \frac{q_{min}}{q_{avr}} \qquad ...(8.9)$$

where,

EU = design emission uniformity (%)

C_v = manufacturing coefficient of variation for point or line-source emitters

n = number of emitter per plant or value of n is 1 for line-source emitters

q_{min} = minimum emitter discharge rate at minimum pressure in the section (lph)

q_{avr} = average or design emitter discharge rate (lph)

8.3.4 Uniformity Coefficient

The degree of emitter flow variation can be represented by a term called uniformity coefficient as defined by Christiansen (1942). The uniformity coefficient for emitter flow variation can be expressed as

$$CU = (1 - \frac{\sum |x|}{m \times n}) \times 100 \qquad ...(8.10)$$

where,

> CU = uniformity coefficient (%)
>
> x = absolute deviation of the individual observation from the mean discharge (lph)
>
> m = mean of all observations (lph)
>
> n = number of observations

8.4 Water Application Uniformity: Statistical Uniformity

The water application uniformity is affected by hydraulic design, land slope, operating pressure, pipe size, emitter spacing and emitter discharge variability. The coefficient of variation and the statistical uniformity are used to evaluate emitter discharge variation and to differentiate between hydraulic design and emitter performance variation. Statistical uniformity is used to evaluate water application uniformity within a sub-main unit or throughout the drip irrigation system. Statistical uniformity of the emitter discharge rate is determined as follows:

$$U_s = (1 - C_v) \times 100 \qquad \qquad ...(8.11)$$

where,

> U_s = statistical uniformity of the emitter discharge rate
>
> C_v = coefficient of variation

Table 8.2 shows the statistical uniformity and its equivalent emission uniformity, as estimated using the lower quartile and acceptability of the design. Table 8.3 provides criteria for accepting the design emission uniformity, unless economic consideration dictates higher or lower emission uniformity (IS: 10799).

Table 8.2: Comparison of statistical uniformity and emission uniformity

Method acceptability	*Statistical uniformity, U_s (%)*	*Emission uniformity, EU (%)*
Excellent	100 – 95	100 – 94
Good	90 – 85	87 – 81
Fair	80 – 75	75 – 68
Poor	70 – 65	62 – 56
Unacceptable	<60	<50

Table 8.3: Recommended ranges of design emission uniformity (EU)

Emitter type	Spacing (m)	Topography	Slope (%)	EU range (%)
Point source on perennial crops	> 4	Uniform steep or undulating	< 2	90 – 95
			> 2	85 – 90
Point source on perennial or semi-permanent crops	< 4	Uniform steep or undulating	< 2	85 – 90
			> 2	80 – 90
Line source on annual or perennial crops	All	Uniform steep or undulating	< 2	80 – 90
			> 2	70 – 85

Example 8.1. Estimate the statistical uniformity (U_s) of emitters for a drip irrigation sub-main unit with the following field data.

Given: Time (seconds) required to fill a 100 ml container from 18 individual emitters are 64, 79, 67, 71, 75, 81, 68, 85, 75, 69, 85, 77, 89, 68, 81, 90, 65, and 61.

Sol.: Individual emitter discharge (ml/s) : 1.56 , 1.26, 1.49, 1.41, 1.33, 1.23, 1.47, 1.17, 1.33, 1.45, 1.17, 1.30, 1.12, 1.47, 1.23, 1.11, 1.53, and 1.64

Average emitter flow $q(\bar{x})$ = Total emitter discharge/number of emitters

$$= 24.27/18 = 1.35 \text{ ml/sec}$$

Standard deviation, $\sigma = 0.16$ ml/sec

Coefficient of variation, $C_v = \dfrac{\sigma}{\bar{x}} \times 100$

$$= \dfrac{0.16}{1.35} \times 100 = 11.85$$

Statistical uniformity,

$$U_s = 100 \, (1 - C_v)$$
$$= 100 \, (1 - 0.11)$$
$$= 89\%$$

Therefore, the emitters are in the good category.

Example 8.2. In a mango orchard, sample emitter discharges were collected from 10 points along a lateral line. The rated discharge capacity of emitter is 4 lph. The discharges collected are given as 3.75, 3.60, 3.85, 4.00, 4.12, 3.95, 3.40, 4.20, 3.20, and 3.80. Estimate the coefficient of uniformity and emission uniformity.

Sol.: The average of emitter discharges is 3.78 lph.

Absolute deviation of the individual observation from the average discharge is 2.41 lph. The average of lowest one-fourth samples of collected discharge is 3.40 lph. The coefficient of uniformity can be determined by,

$$CU = (1 - \frac{\Sigma|x|}{m \times n}) \times 100$$

$$= (1 - \frac{2.41}{3.78 \times 10}) \times 100$$

$$= 93.62 = \mathbf{94\%}$$

and emission uniformity is determined as,

$$EU = \frac{q_n}{q_{avr}} \times 100$$

$$= \frac{3.40}{3.78} \times 100$$

$$= 89.44 = \mathbf{90\%}$$

Ans. 94%, 90%.

Example 8.3. In order to check the uniform application of water through dripper, discharge from 15 drippers were measured for 15 minutes from the field. Determine the average discharge of dripper, coefficient of variation and statistical uniformity of dripper discharge rate.

Dripper no.	1	2	3	4	5	6	7	8	9	10	11	12	13	14	15
Discharge (ltr)	2.7	2.8	2.6	2.5	2.8	2.4	2.9	2.3	2.1	2.6	2.0	2.1	2.0	2.9	2.4

Sol.: As the measured discharge is for only 15 minutes duration, these will be multiplied by 4 to get discharge in lph. Below are calculated values.

No.	1	2	3	4	5	6	7	8	9	10	11	12	13	14	15
L	1.8	2.1	1.6	2.3	2	2.4	2.2	2.3	2.1	2.3	2	2.1	2	2.4	2.1
lph	7.2	9.2	6.4	9.2	8	9.6	8.8	9.2	10	6.8	8	7.2	8	9.6	6.4

Average discharge =

$$\frac{7.2+9.2+6.4+9.2+8+9.6+8.8+9.2+10+6.8+8+7.2+8+9.6+6.4}{15} = 8.24 \text{ lph}$$

Coefficient of variation, $C_v = \frac{\text{Standard deviation}}{\text{Average flow}} \times 100$

Standard deviation, $SD = \sqrt{\frac{\sum_{i=1}^{n}(x_i - \bar{x})^2}{n-1}}$

where, x_i is the individual dripper discharge (lph), $\bar{x}$ is the average dripper discharge and n is the number of observations. The calculations are shown below.

No.	1	2	3	4	5	6	7	8	9	10	11	12	13	14	15
L	1.8	2.3	1.6	2.3	2	2.4	2.2	2.3	2.5	1.7	2	1.8	2	2.4	1.6
lph	7.2	9.2	6.4	9.2	8	9.6	8.8	9.2	10	6.8	8	7.2	8	9.6	6.4
$(x_i - \bar{x})^2$	1.56	0.56	4.20	0.56	0.20	1.32	0.12	0.56	2.40	2.72	0.20	1.56	0.20	1.32	4.20

Therefore,

$$\text{Standard deviation}, SD = \sqrt{\frac{\sum_{1=1}^{n}(x_1 - \bar{x})^2}{n-1}} = \sqrt{\frac{21.72}{15-1}}$$

Coefficient of variation,

$$C_v = \frac{\text{Standard deviation}}{\text{Average flow}} \times 100 = \frac{1.24}{8.24} \times 100 = 15\%$$

We know that statistical uniformity, $U_s = (1 - C_v) \times 100 = (1 - 0.15) \times 100 = 85\%$ which comes under good as far as acceptability is concerned.

Ans. 8.24 lph, 15%, 85%

9

INSTALLATION, MANAGEMENT AND MAINTENANCE OF DRIP IRRIGATION SYSTEMS

Improper maintenance of the drip irrigation system can cause failure of the properly designed systems. The sensitivity of emitter to clogging, network of lateral pipes, filtration units and fertigation systems add the complexities in the systems and need careful maintenance. You can maintain a system, if it is appropriately installed. Proper installation can ease your maintenance job. Let us first discuss about installation of drip irrigation system.

9.1 Installation

We assumed that a source of water already exists. Installation of the drip irrigation system can be divided into three stages: (i) Fitting of filter units, (ii) laying of mains and sub-mains, and (iii) laying of laterals with emitters.

9.1.1 Installation of Filter Unit

The following points should be considered for deciding the position of filter unit.

- There should be minimum use of fittings such as elbows and bends.
- The filter unit should be located on the delivery side of the pump.
- The filter size should be in accordance with the capacity of the system.

9.1.2 Installation of Mains and Sub-mains

Except for a fully portable system, both mains and sub-mains are installed underground at a minimum depth of about 50–60 cm, such that they are unaffected by cultivation or by heavy agricultural machinery. Before laying PVC pipes in a trench, it may be padded with sand to avoid the expansion and contraction of compacted soils. Even for systems which have portable laterals that are removed at the end of each season, it is common practice to install permanent underground sub-mains. Generally, sub-mains run across the direction of the rows.

9.1.3 Laying of Laterals

Generally, laterals are laid on the ground surface and placed along contours on sloping field. Laterals laid from the reels should be left on the ground for 5–6 hours so that it can release the twisting formed in the reel. The downstream end of the lateral can be closed by simply folding back the pipe and closing it with an end plug. This can be easily slipped for flushing. The simplest connection for low-pressure system is for the lateral to be inserted directly into the sub-main. Cutting a slightly undersized hole in the sub-main with either a wood auger bit or a metal twist drill, is desirable. The hole is expanded with the tapered tool, and then the lateral is inserted quickly after withdrawing the taper. The lateral is cut at angle of about 45° at the end. When deciding the lateral length for installation, take into account that the lateral shrinks 1 meter per 100 meters. Holes are made using punch on the lateral as per the required spacing. The on-line point-source emitters should be fixed such that outlets facing upward after laterals are spread on the ground.

9.2 Testing of the System

Inspection at regular interval will ensure the proper functioning of the drip systems. This can be done weekly, monthly or on seasonal basis, keeping in view the water quality and attributes of the system components. Pump efficiency need to be tested at least once in 3 years. Higher pump efficiency, more than 80%, is desirable, as it will also save the ever increasing energy cost. All the gate valves and flush valves should be opened before testing. The filter should be backwashed till clean water comes out through its flush valve. The flush valve should be closed after flushing of sub-main. The working of air release valve is checked at the sub-main. The pressure at inlet and outlet of the filter should be obtained by regulating the bye-pass valve. Discharges from emitters are measured at normal operating pressure and their discharge variation and emission uniformities are determined.

9.3 Maintenance of Drip Irrigation System

The first step in maintaining a drip irrigation system is to take the sample of the water from the source which is going to be used. This will help in determining the flushing requirement of the drip system. There are three types of clogging problems that need to be taken care of while preparing the maintenance program. These are physical, chemical and biological. Sand and silt may cause the physical hazards to the system. Appropriate filtration technique is required to remove particles above the maximum allowable particle size for the emitters being used.

Chemical clogging hazards include the precipitation of excessive salt concentration. Irrigation water containing more than 0.1 ppm sulphides can encourage the growth of sulphur bacteria within the irrigation system. Biological clogging hazards result from the growth of bacterial slimes and algae within drip lines and emitters. They combine with clay particles to block the emitters. Periodic preventive maintenance is required for successful operation of a drip irrigation system. The general maintenance consists of following steps.

(i) The emitter functioning, wetting pattern and leakage of pipes, valves, and fittings should be checked regularly.

(ii) The placement of emitters should be ascertained. If the placement is disturbed, put the emitters in proper location.

(iii) Leakages through filter gaskets in the lids, flushing valves and fittings, etc. should be monitored regularly.

9.3.1 Flushing the Mainline, Sub-main and Laterals

During the installation process and before connecting the drip laterals, flush the main and sub-mainlines. In case of failure or burst in the main or sub-main pipe, it is recommended to flush the dirt from lines by releasing the ends of the main and sub-mainlines and flushing them for at least 10 minutes. Filters usually do not prevent the entry of suspended particles and water soluble matter. Gradually over the time, a layer of sediment or organic matter settles in the mainline and drip laterals. The end of the lateral line experiences the more accumulation of sediment and therefore, it is recommended to flush the lines after 2–4 weeks of initial operation. Normally, flushing 3–4 times per season when irrigating with good quality water (shallow/deep tube well), and at least once every fourth operation when irrigating with dirty water (ponds/canal) is suggested.

During the crop growing period, daily checkup of screen/media filter performance and pressure and flow rate of the system should be carried out to meet the designed performance. Weekly observations on emitters clogging, cleaning of screens and strainers can be made. Care must be taken on monthly basis regarding the maintenance requirement of screen/media filter, comparison of lateral pressure to designed pressure and measuring of emitter discharge rate at selected locations.

9.3.2 Filter Cleaning

Filter is the core of a drip irrigation system and its failure will lead to clogging of the entire system. Keeping it serviceable and clean will prolong the life of the irrigation system. Pressure differential across the filter is the correct indication of the timing of cleaning of the filter. Some of the steps are outlined here to maintain the filters.

1. Examine screens and seams to ensure that they are undamaged.

2. Ensure proper tightness between the screen, cover and body of the filter, in order to prevent lateral water leaks. If using a disc type filter, ensure the disc assembly is properly tightened.

3. Working with a partially clogged filter is detrimental to both the filter and the quality of filtration. When the pressure differential reaches approximately 5 meters, the system should be flushed and cleaned.

4. Avoid entry of dirt into the pipe when the filter is opened for cleaning.

5. Flush the filter before irrigation operation starts. If the filter is very dirty, clean more frequently. Automatic back-flushing may be appropriate.

The cleaning procedures for various filters are given below.

9.3.2.1 Media Filter

The filter should be backwashed everyday for 5 minutes to remove the silt and other dirt accumulated during the previous day's irrigation. Once in a week, while backwashing, the backwash water should be allowed to pass through the lid instead of the backwash valves. The sand in the filter bed is stirred up to the filter candles without damaging them. Whatever dirt is accumulated deep inside the sand bed will get free and goes out with the water through the lid. For removal of aggregates and algae, fill the container with chlorine and allowed to soak for 24 hrs. After rinsing, pour off the water and leave the filter in a dry and ventilated place. Greasing and repair is done to avoid rusting as regular maintenance.

9.3.2.2 Screen Filter

The screens and sieves are needed to be examined to ensure that they are undamaged. In order to prevent lateral water leakage, ensure proper tightening of screens, cover and body of the filter. Before the start of a drip irrigation system, the flushing valve on the filter link is opened to flush out the dirt and silt. The filter element is taken out from the filter and it is cleaned in flowing water. The rubber seals are taken out from both the sides and precaution should be taken while replacing the rubber seals, otherwise they may get damaged.

9.3.3 Chemical Treatment

When normal flushing is not sufficient to clean the system and bring uniform emission, chemical treatment is recommended. Clogging or plugging of emitters and orifices of bi-wall is due to precipitation and accumulation of certain dissolved salts like carbonates, bicarbonates, iron, calcium and manganese salts. The clogging is also due to the presence of microorganisms and the related iron and sulphur slimes due to algae and bacteria. The clogging of emitters is usually removed by chemical treatment. Chemical treatment commonly used in drip irrigation systems includes addition of chlorine and/or acid to the water supply. The frequency of chemical treatment is decided on the degree of clogging and quality of water. As a general rule, acid treatment is performed once in 10 days and chlorine treatment once in 15 days.

9.3.3.1 Acid Treatment

Flushing with acid solution can dissolve the precipitation caused by dissolved salts. Hydrochloric acid (30%) is injected into the drip system at the rate suggested in the water quality analysis report. All the sub-mains and laterals are flushed before the acid treatment and check the discharge of the system before treatment. A solution equal to one-tenth (1/10) of the maximum discharge of the injector pump is prepared and treatment is performed till a pH of 4 is observed at the furthest lateral. After that irrigation is continued for 30 to 60 minutes.

9.3.3.2 Chlorine Treatment

Oxidation treatment is applied to decompose the organic matter and preventing the building up of algae and bacteria. Chlorine is an oxidizing agent, which kills bacteria and other organic matter and prevents build up. The most common chlorine compounds are sodium hypochlorite (liquid) and calcium hypochlorite (solid). There are three main chlorine application methods:

1. Continuous treatment
2. Superchlorination
3. Intermittent treatment

1. **Continuous Treatment:** It is the simplest method of removing blockage problems in drip irrigation system. The water is treated continuously to keep the system clean from organic material. The required concentration varies according to water quality. Usually, the injection rate is adjusted such that 1 ppm of chlorine is detectable at the end of the furthest lateral from the pump. The common concentration to begin with is usually 5–10 ppm (mg/L).

2. **Superchlorination:** When the irrigation system has operated for some time and there is an accumulation of organic matter inside the pipeline, superchlorination is needed to dissolve the organic slime. Chlorine is normally injected at a concentration of 500 ppm. The system is shut down and left for 24 hrs. The system is then flushed step-by-step, first main line, then sub-main, and finally through the end of the laterals.

3. **Intermittent Treatment:** Although the filter system operates properly, microorganisms such as algae, fungi and bacteria may pass through it into the irrigation system. To prevent a build-up of this organic matter in the laterals, periodic sterilization by chlorination is needed. During the last hour of irrigation, chlorine concentration of 10–20 ppm is applied for 30 minutes in the irrigation water. The chlorine is left in the pipe until the next irrigation.

The rate of injection of liquid chlorine or acid depends on the system flow rate and can be determined by

$$q_c = K \frac{uQ_s}{c'} \qquad \qquad ...(9.1)$$

where,

q_c = rate of injection of the chemical into the system (lph)

K = conversion constant, 3.6×10^{-3} for metric units

u = desired dosage in irrigation water (ppm)

Q_s = irrigation system discharge capacity (lps)

c' = concentration of the desired component in liquid chemical concentrate (kg/L)

Chlorination should be used regularly to kill algae and loosen bonded organic matter within the system for flushing out. It has no effect on scale (lime or calcium carbonate) deposits in the system. Liquid sodium hydrochloride is the recommended chlorine source. It is injected into the system during regular operation usually using the fertigation equipment. Rates of 5 to 20 ppm chlorine may be required, depending on the severity of the problem. It is essential to measure the active chlorine in the irrigation water at the start and end of the system during chlorination to ensure that the required concentrations are reached. Chlorine is more effective in acid waters. High pH or alkaline waters should be acidified to a pH of 6.5 for effective chlorine treatment. The Eqn. 9.1 may be given in a different form with easier units as follows.

$$q = \frac{C_1 \times Q}{C_0 \times 10} \qquad \qquad ...(9.2)$$

where

q = discharge of injected chlorine solution (lph)

C_0 = percentage of active chlorine in injected solution

C_1 = desirable concentration of active chlorine in irrigation water (mg/L)

Q = discharge of the treated system (m^3/h)

Example 9.1. In drip irrigated vegetable plot, the system flow rate is 125 m^3/h. Active chlorine percentage in the solution was 10% and the desired concentration is 15 mg/L. Estimate the discharge of injected chlorine solution.

Sol.: We can apply Eqn. 9.2 and find the result. Applying the equation, we get

$$q = \frac{15 \times 125}{10 \times 10} = \textbf{18.75 lph}$$

Ans. 18.75 lph.

10

SALT MOVEMENT UNDER DRIP IRRIGATION SYSTEMS

Irrigation requires relatively large quantities of water. Appropriate selection of crops, irrigation methods and management practices can enable us to utilize wide range of irrigation water quality. Non-conventional water resources such as saline drainage water, brackish groundwater and treated waste water are alternatives to freshwater resources. However, careful management is required to safeguard the environment and promotes sustainable agricultural production without degrading our land and water resources. In conventional surface irrigation systems, the whole field is flooded with irrigation water and usually the water flow through the soil is one-dimensional. Excess water drains out of the root zone and helps in leaching of salts out of the root zone.

In case of drip irrigation systems, when emitters are placed above the soil surface, two-dimensional water flows occurs within the soil where only portion of the soil surface is wetted. Similarly, three-dimensional water and salt movement occurs within the root zone where the emitters are placed beneath the soil surface as in case of sub-surface drip irrigation. Under these conditions, excess water does not flow into deep drainage and also does not allow leaching the salt below the root zone. For example, some of the water moving from a buried drip irrigation emitter will move laterally or up towards the soil surface. When water arrives at the soil surface, it is evaporated, and the residual salt accumulates on the surface which may be leached back into the root zone by subsequent rainfall (Fig.10.1). It has been observed that in arid and semi-arid regions, the salt accumulation in the root is always a concern for farmers irrigating crops with drip irrigation systems. Excess sodium (Na^+) can lead to breakdown of clay particle structure and can clog the soil and reduce infiltration rate to nearly zero.

Under conditions found in commercial fields, irrigation with saline water (ECW of 1.5 to 2.0 dS/m) results in relatively low salinity levels in the area extending downward from surface drip lines and larger salt accumulations in the areas between drip lines and near the edge of the bed for sandy loam and clay loam (Figs. 10.1 A and B). The low-salinity zone extends further horizontally in the clay loam soil than in the sandy loam. Under sub-surface drip irrigation using saline water ($EC_W = 2.5$ dS/m), salt normally accumulates above the drip line, with the highest levels occurring near the soil surface (Fig. 10.2). Soil salinity is found lower around and below the drip line (Hanson and May, 2011).

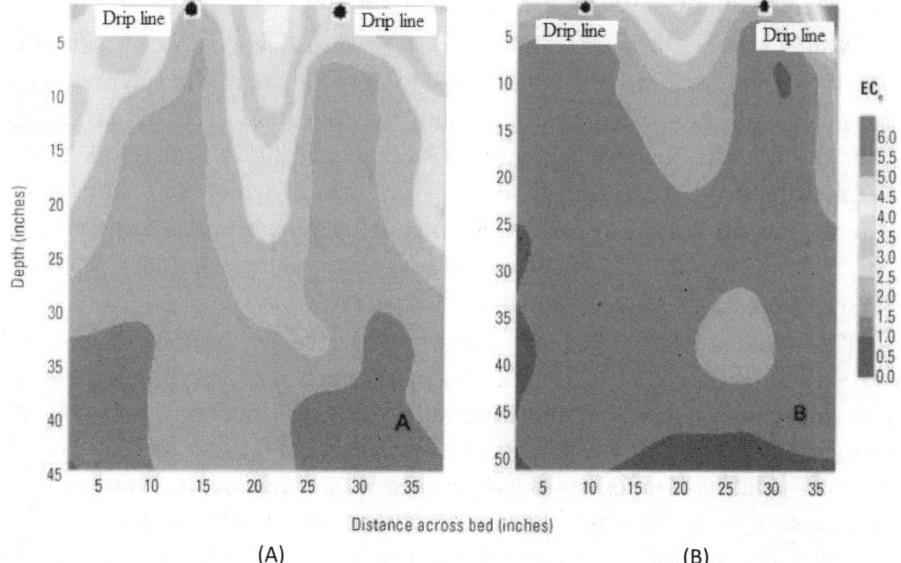

Fig. 10.1: Salt distribution (EC_e) around surface drip lines for (A) sandy loam soil and (B) clay loam soil. Units are dS/m

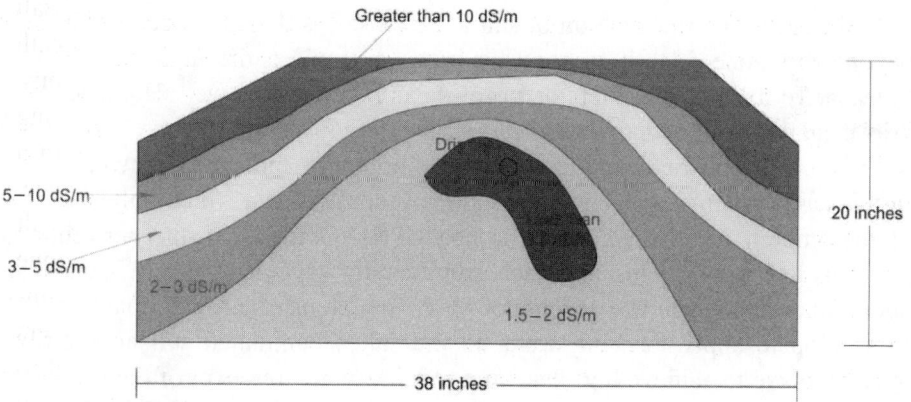

Fig. 10.2: Salt distribution around a sub-surface drip line. Units are dS/m

Surface evaporation and net flux of water across the soil surface under drip irrigation is spatially variable. At and near the dripper, the net water flux will likely be downwards, but further away evaporative fluxes will exceed infiltration, especially during dry periods, leading to an upward flux of water. The use of surface mulches which reduce evaporative fluxes can have a large impact of the direction and magnitude of vertical water and salt flux. It has been reported that irrigation during the daytime, produces different soil-water distributions to irrigations conducted at night time due to differences in upward flux. Thus, at the end of a dry summer, during which a crop has been drip irrigated, salt patterns are likely to be highly variable.

Winter rain could leach salt, but may be insufficient to leach salt from areas of high concentration. In some cases, rainfall may mobilize salt previously accumulated on the soil surface back into the root zone creating an adverse impact on root zone osmotic potentials (Raine et al., 2005). Cote et al. (2003) simulated the flow of a pulse of solutes from drip irrigation and showed that solute applied at the end of the irrigation ends up deeper in the soil compared to when it was applied at the start of the irrigation, owing to an increase in the ratio of downward to lateral water flux over time.

This is completely different to what would happen for one-dimensional flow. Such studies suggest that much more research is required to understand solute transport in drip systems especially over an irrigation cycle and the interaction with rainfall events. Plant roots also play a major role in soil-water and solute dynamics by modifying the water and solute uptake patterns in the rooting zone. Also the implementation of drip irrigation systems raises the vital issues in relation to soil-water and salt management. The adoption of drip irrigation technologies and practices can result in root zone salinisation problems. The leaching calculated using one-dimensional equation may not adequately describe leaching under some field conditions.

Normally, the reclamation of salt affected soil is done by leaching the salts beyond root zone. Idea is to pond the water over the entire field and allows it to leach. To do this, sprinklers or furrows can be used very well. However, with drip irrigation, salt accumulation takes place mainly along the rows as visible in Fig.10.1. Therefore, leaching water need not to be applied to the whole field. In this case, water is applied to one-third of the field area which only one-third of the amount of leaching water is needed as compared to the conventional leaching techniques. This approach significantly saves the water. Four to six lateral lines are placed at a spacing of 40–50 cm along one row of plants mainly fruits crop to apply leaching water to the salt accumulated soil area. Efforts must be made to achieve high leaching efficiency, i.e. removal of the maximum salt per unit of leaching water. Leaching water is normally applied for 24 hours and stopped for 3–4 days and started for another 24 hours. Soil samples may be collected after 4–5 days of application of last leaching water to check the salinity level in crop root zone.

10.1 Salinity Measurement

Salts in saline soils originate from the natural weathering of minerals or from fossil salt deposits left from ancient seabeds. On arid or semi-arid regions, salts accumulate when the irrigation water or groundwater seepage evaporates. Examples of common salt compounds are sodium chloride (NaCl), sodium sulfate (Na_2SO_4), magnesium sulfate ($MgSO_4$), calcium sulfate ($CaSO_4$), and calcium carbonate ($CaCO_3$). An electrical current is passed through a soil solution extract to measure salinity of the soil. The ability of the soil solution extract to carry a current, is called electrical conductivity (EC). EC is measured in deciSiemens per meter (dS/m), which is the numerical equivalent to the millimhos per centimeter. The lower the salt content of the soil, the lower the dS/m rating and the less the effect on plant growth. Yields of most crops are not significantly affected where salt levels are 0 to 2 dS/m. Generally, a level of 2 to 4 dS/m affects some crops. EC is related with total cations and osmotic pressure of soil water extract, the following expressions can be used to evaluate the salt concentrations.

Salt concentration (ppm) $= 640 \times$ EC (millimhos/cm) ...(10.1)

Total cation concentration (mEq/L) $= 10 \times$ EC (millimhos/cm) ...(10.2)

Osmotic pressure (atm.) $= 0.36 \times$ EC (millimhos/cm) ...(10.3)

We know that 1 ppm is equal to 1 mg/L and if ppm is divided by the equivalent weight, we get mEq/L.

Example 10.1. It is given that electrical conductivity of the soil water extract is 8 millimhos/cm. Convert it into ppm, micromhos, and mhos/cm.

Sol.: Electrical conductivity (EC) of the soil water extract

$$= 8 \text{ millimhos/cm}$$

$$= 8/100 = 0.08 \text{ mhos/cm}$$

$$= 8 \times 1000 = 8000 \text{ micromhos/cm}$$

As we know that

Salt concentration (ppm) $= 640 \times$ EC (millimhos/cm)

$$= 640 \times 8$$

$$= 5120 \text{ ppm}$$

Ans. 5120 ppm

Example 10.2. What is the concentration of salts (mg/L) in water with EC = 2.4 dS/m?

Sol.: Salt concentration $=$ EC $\times$ 640

$$= 2.4 \times 640 = 1500 \text{ mg/L}$$

Ans. 1500 mg/L

10.2 Leaching Requirements

Ragab (2002) reported that soil salinity does not reduce crop yield significantly until a threshold level is exceeded. To avoid yield loss when salt concentrations exceed their tolerance limits, excess salts must be leached below the root zone. Thus, when the net depth of applied water is calculated for scheduling, an additional depth of water based on the salinity level should be added for leaching. The leaching requirement (*LR*) is usually calculated as:

$$LR = \frac{D_d}{D_i} = \frac{C_i}{C_d} \qquad \qquad ...(10.4)$$

where,

D_d = depth of water passing below the root zone as drainage water

D_i = depth of applied irrigation and rainfall water

C_d = salt concentration of the drainage water above which yield reduction occurs

C_i = salt concentration of the irrigation water

The initial equation developed by (Reeve et al., 1967) is given as

$$LR = \frac{EC_W}{EC_D} = \frac{C_i}{C_d} = \frac{L_N}{D_N + L_N} \qquad \qquad ...(10.5)$$

where

EC_W = electrical conductivity of irrigation water

EC_D = electrical conductivity for water draining under the root zone

LR = net annual leaching requirement (mm)

D_N = net annual depth of irrigation to meet consumptive use of water (mm)

Bernstein (1973) reported the following equations to determine the leaching requirement.

$$LR = \frac{EC_W}{(5EC_E - EC_W)} \qquad \qquad ...(10.6)$$

but to measure leaching requirement under drip irrigated crops, the following equation was suggested.

$$LR = \frac{EC_W}{2 \times \max EC_E} \qquad \qquad ...(10.7)$$

where, EC_E is electrical conductivity corresponding to 100% yield loss.

Example 10.3. Determine the leachate salinity. Irrigation water salinity $(EC_w) = 2$ dS/m. Applied water depth $(D_i) = 1250$ mm/season. There is no precipitation during the growing season. Crop water demand $(ET_c) = 1,000$ mm/season. Average soil moisture content is the same at the beginning and end of the growing season.

Sol.: We know the leaching requirement as

$$LR = \frac{D_d}{D_i} = \frac{1250 - 1000}{1250} = 0.20$$

We know that $LR = \dfrac{EC_w}{EC_D}$

Therefore, $\quad EC_d = \dfrac{EC_W}{LR} = \dfrac{2}{0.2} = 10$ dS/m

Ans. 10 dS/m

10.3 Salt Tolerance of the Crop and Yield

Under drip irrigation system, frequent application of water is provided which keeps the salt concentration in the soil to a minimum level. Frequent irrigation usually keeps the salt concentration in the soil water almost equal to the salt concentration in irrigation water. But if irrigation interval is increased, the salt accumulation can take place as the root zone soil dries. It has been observed that with poor quality irrigation water, crop yield under drip irrigation systems is usually higher than that of other methods of irrigation with the same quality of irrigation water. This is mainly attributed to the fact that salts remain diluted by the continuous supply of irrigation water.

Information of electrical conductivity of irrigation water and saturated soil extract are essential for salinity management. To determine the relative yield, maximum electrical conductivity of saturated soil extract is necessary. This is something at which yield of the crop will be zero. The maximum electrical conductivity of saturated soil extract (EC_E) for several crops Table 10.1 (Keller and Bliesner, 1990). If the electrical conductivity of irrigation water (EC_W) is lying between the minimum EC_E and (max EC_E + min EC_E)/2, then relative crop yield (Y_R) can be obtained from the following expression.

$$Y_R = \frac{\max EC_E - EC_W}{\max EC_E - \min EC_E} \qquad\qquad ...(10.8)$$

where,

    min EC_E = electrical conductivity of saturated soil extract that will not decrease crop yield, dS/m

    max EC_E = electrical conductivity of saturated soil extract that will decrease crop yield to zero, dS/m

Example 10.4. A well developed lemon orchard under drip irrigation is located in semi-arid region of India. If the electrical conductivity of irrigation water is 3.2 dS/m and net annual consumptive use deficit is 460 mm, determine the net annual leaching requirement and effect of salinity of irrigation water on yield of lemon.

Sol.: From the Table 10.1, we find minimum and maximum values of electrical conductivity of saturated soil extract (EC_E) which is 1.7 and 8. Leaching requirement under drip irrigation is obtained as follows.

$$LR = \frac{EC_W}{2 \times \max EC_E}$$

$$= \frac{3 \times 2}{2 \times 8} = 0.20 \text{ or } 20\%$$

Now take Eqn. 10.5

$$LR = \frac{L_N}{D_N + L_N}$$

or $$LN = R \times (D_N + L_N)$$

or $$L_N = \frac{LR \times D_N}{1 - LR}$$

$$= \frac{0.20 \times 460}{1 - 0.20}$$

$$= \mathbf{115 \ mm}$$

The relative yield as compared to the full potential yield due to salinity of irrigation water can be estimated by

$$Y_R = \frac{\max EC_E - EC_W}{\max EC_E - \min EC_E}$$

$$= \frac{8.0 - 3.2}{8.0 - 1.7}$$

$$= \mathbf{0.76}$$

This means that if full potential of yield is 100 t, then with this salinity level of irrigation water, 76 t of produce will be obtained.

Example 10.5. Under a drip irrigated plot, the EC of the saturated soil paste was recorded as 9 millimhos/cm and EC corresponding to 100% yield loss was 13 millimhos/cm. Find out the leaching requirement.

Sol.: We will apply Eqn. 10.7,

$$LR = \frac{EC_W}{2 \times \max\ EC_E}$$
$$= \frac{9}{2 \times 13} \times 100$$
$$= 34.62 = \textbf{35\%}$$

Ans. 35%

10.4 SALTMED Model

There are many single-process oriented model such as (*i*) models for infiltration (*ii*) models for root water uptake (*iii*) models for leaching or water and solute transport or (*iv*) models for specific applications, i.e. certain irrigation system, soils, region or a crop. The need for comprehensive generic models that account for different crops, water and field management practices was addressed by developing the SALTMED model (Ragab, 2002). The model employs well established water and solute transport, evapotranspiration and crop water uptake equations and successfully illustrated the effect of the irrigation system, the soil type and irrigation salinity level on soil moisture and salinity distribution, leaching requirements, and crop yield in all cases. The model successfully predicted the impact of salinity on yield, water uptake, soil moisture and salinity distribution. The model provides academics with a research tool and field managers with a powerful tool to manage their water, crop and soil in an effective way in order to save water and protect the environment.

Table 10.1: Minimum and maximum values of electrical conductivity of saturated soil extract (EC_E)

Crops	EC_E		Crops	EC_E	
	Min	Max		Min	Max
Field crops					
Cotton	7.7	27	Corn	1.7	10
Sugarbeet	7.0	24	Flax	1.7	10
Sorghum	6.8	13	Broadbeans	1.5	12
Soybean	5.0	10	Cowpeas	1.3	8.5
Sugarcane	1.7	19	Beans	1.0	6.5
Fruits and nut crops					
Date palm	4.0	32	Apricot	1.6	6
Fig, olive	2.7	14	Grape	1.5	12
Pomegranate	2.7	14	Almond	1.5	7
Grapefruit	1.8	8	Plum	1.5	7
Orange	1.7	8	Blackberry	1.5	6

Contd.

Contd.

Crops	EC_E		Crops	EC_E	
	Min	Max		Min	Max
Lemon	1.7	8	Boysenberry	1.5	6
Apple, pear	1.7	8	Avocado	1.3	6
Walnut	1.7	8	Raspberry	1.0	5.5
Peach	1.7	6.5	Strawberry	1.0	4.0
Vegetable crops					
Beets	4.0	15	Pepper	1.5	8.5
Broccoli	2.8	13.5	Lettuce	1.3	8.0
Tomato	2.5	12.5	Radish	1.2	9.0
Cucumber	2.5	10	Onion	1.2	7.5
Spinach	2.0	15	Carrot	1.0	8.0
Cabbage	1.8	12	Beans	1.0	6.5
Potato	1.7	10	Turnip	0.9	12.0

10.4.1 Equations of the Model

The SALTMED model addresses the important processes such as evapotranspiration, plant water uptake, water and solute transport under different irrigation systems including drip irrigation, drainage and the interrelationship of crop yield and water use. The evapotranspiration has been estimated by using the Modified Penman-Monteith equation. In case, the weather data such as temperature, radiation and, wind speed, etc. are not available, the model can use pan evaporation data to calculate the evapotranspiration. The effective rainfall is defined as the part of the rainfall that is available for infiltration through the soil surface. The model estimates it in three ways: (*i*) as a percentage of total rainfall, (*ii*) calculate according to the FAO-56 (1998) procedure, and (*iii*) taken to be equal to total rainfall. The water uptake rate formula suggested by Cardon and Letey (1992a), which determines the water uptake S per day as given below:

$$S = \left[\frac{S_{MAX}}{1 + \left(\frac{a.h + \pi}{\pi_{50}} \right)^3} \right] . \lambda \qquad ...(10.9)$$

where,

$$\lambda(z) = 5/3L \text{ for } z \leq 0.21 \qquad ...(10.10)$$

$$\lambda(z) = 25 / 12L^* (1 - z/L) \text{ for } 0.2L < z \le L \qquad \text{...(10.11)}$$

$$\lambda(z) = 0.0 \text{ for } z > L \qquad \text{...(10.12)}$$

where, S_{MAX} is the maximum potential root water uptake in time t; z is the vertical depth taken positive downwards, λ is the depth and time dependent fraction of total root mass, L is the maximum rooting depth, h is the matric pressure head, π is the osmotic pressure head; π_{50} is the time dependent value of the osmotic pressure at which S_{MAX} is reduced by 50%, and a is a weighing coefficient that accounts for the differential response of a crop to matric and solute pressure. The coefficient a equals π_{50}/h_{50}, where h_{50} is the matric pressure at which S_{MAX} is reduced by 50%. The maximum water uptake, S_{MAX}, is determined as

$$S_{MAX} = E_{T0} \times K_{CB} \qquad \text{...(10.13)}$$

where, K_{CB}, is the crop transpiration coefficient or basal crop coefficient. The values of K_{CB} and K_C are available in the model's database. These data can be used in the absence of measured values.

The flow of the water in soils is described by the Richards equation, which is a partial non-linear differential equation, and based on the Darcy's law and Law of mass continuity. The equation of Darcy's law can be given as,

$$q = -K(h) \frac{\delta H}{\delta Z} \qquad \text{...(10.14)}$$

where, q is the water flux, $K(h)$ is the hydraulic conductivity as a function of soil water pressure head, Z is the vertical coordinate directed downwards with its origin at soil surface, and H is the hydraulic head which is the sum of the gravity head, Z, and the pressure head, ψ, thus:

$$H = \psi + Z \qquad \text{...(10.15)}$$

The Richards type equation for vertical transient-state flow water in a stable and uniform segment of the root zone is expressed as below.

$$\frac{\partial \theta}{\partial t} = -\frac{\partial}{\partial z}\left[K(\theta) \frac{\partial(\psi + z)}{\partial z} \right] - S_w \qquad \text{...(10.16)}$$

where, θ is volume wetness; t is the time; z is the depth; $K(\theta)$ is the hydraulic conductivity; ψ is the matrix suction head; and S_w is the extraction by plant roots. The movement of solute in the soil system depends greatly on the path of water movement. For detail information, please refer to Ragab (2002). The leaching requirement is calculated in the SALTMED model as a ratio of the salt concentration of the irrigation water to that of the drainage water or the mean salinity level of the root zone as given in Eqn. 10.4.

The input data requirement of the model are plant characteristics for each growth stage, soil characteristics, weather data, and water management data include the date and amount of irrigation water applied and the salinity level of each applied irrigation (Fig. 10.2).

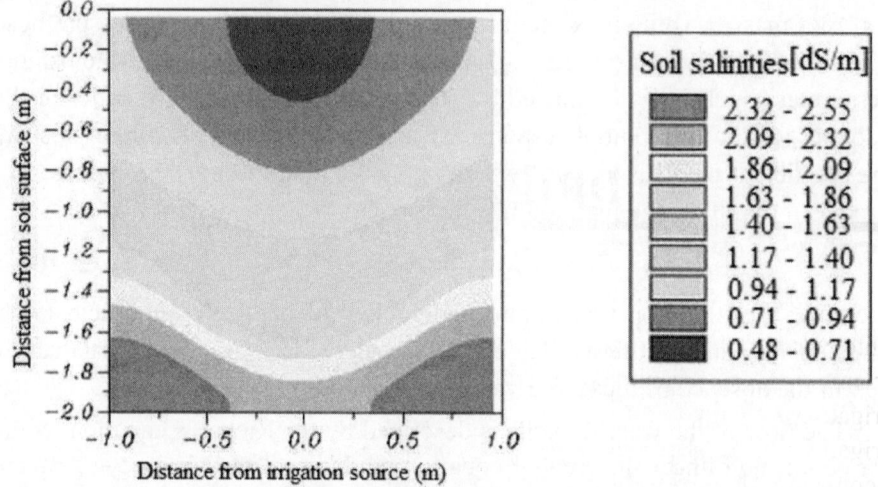

Fig. 10.3: Soil salinity profile under emitter (Ragab, 2002)

A good number of data are available in the databases of the model such as crop database, soils database which contains the hydraulic characteristics and solute transport parameters of more than 40 different types of soils and irrigation system database. The model is able to analysis the effect of the different irrigation systems, soil type, the irrigation salinity level on soil moisture and salinity distribution, leaching requirements, and crop yield. Figure 10.3 indicates that near the emitter, the soil salinity is less.

11

AUTOMATION OF
DRIP IRRIGATION SYSTEMS

Automation of drip irrigation system refers to operation of the system with no or minimum interventions of operators. Automation is most suitable to drip irrigation systems because it supplies controlled and pre-determined amount of irrigation water to irrigate large area from the same source of water. It is always inconvenient to precisely irrigate the farm field without automation. Let us take an example, you may not get labor to operate drip systems frequently and that too for short durations of time, which in many cases is required to maximize yields and avoid wasting of water and other valuable inputs. The problem worsens if it is required to turn valves on for a few seconds after every 30–40 minutes.

It has been realized that controllers and valves are cost effective and reliable, leaving labor to perform other and more important tasks to grow a better crop. Nowadays, relatively inexpensive controllers and valves are available in the market which can be installed in any application with little effort. Controllers are available for applications with or without power, and for situations where simplicity and low cost are the key factors. Solenoid activated valves are available in a multitude of sizes and configurations, and are equally simple and cost effective. Besides above, automation of irrigation systems provides the following advantages.

- It saves our precious water.
- It starts and stops irrigating at the pre-determined or programmed time.
- It can take into account the effective rainfall for scheduling irrigation.
- No need to visit farm at odd times.
- Adequate quantity of water and nutrients supply result in quality produce and high yield.
- No leaching of minerals and nitrogen vital for plants healthy growth.
- Avoid the ill effects of overirrigation which causes development of salinity.

- System can be operated at night, thus the daytime can be utilized for other agricultural activities.

The disadvantages include the expensive systems, needing support and most automated irrigation systems need electricity.

11.1 Types of Automation

Automation systems are classified on the basis of type and scope of control or even time can be the basis of operation. The basic objective is to prepare a schedule based on crop water requirements. First, let us discuss the classification of automation on the basis of type of control. On the basis of type of control, the automated irrigation systems can be classified into two categories, i.e. sequential and non-sequential type. In sequential systems, the field is divided into different smaller units, which is irrigated one at a time in a well laid out sequence, whereas in the non-sequential systems, the smaller units are irrigated randomly based on the crop water needs. Sequential system can be classified as (*i*) hydraulically operated, (*ii*) electrically operated, and (*iii*) combination of above two methods.

11.1.1 Hydraulically Operated Sequential System

The sequential system is particularly suited for irrigation at low discharges through small-diameter tubing. The water metering device and a hydraulic valve are located at each connection to the mainline (Fig.11.1).

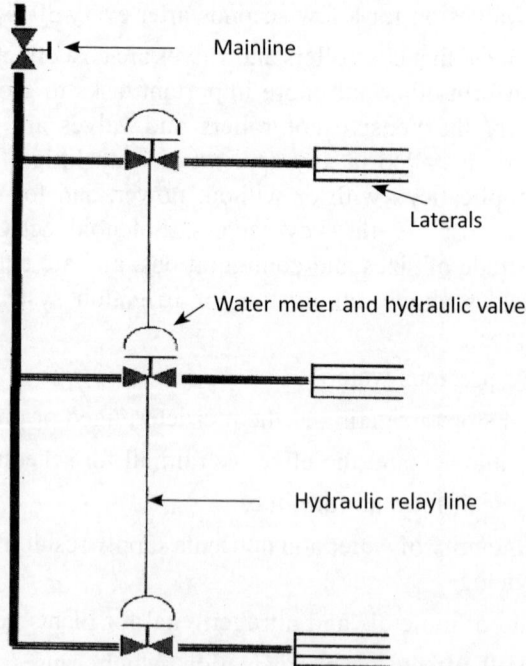

Fig. 11.1: Sequential hydraulic operated automated irrigation system

The metering valves are pre-set for the required amount of water to be applied to the field or sub-plots in the beginning of the irrigation cycle. The amount of water is determined according to crop water requirement. Water under pressure passes through the first metering valve to the hydraulic valve on the mainline and stops the flow of water. The metering valve closes after the pre-determined quantity of water has passed through the first head. The pressure at the head of the hydraulic valve is released, and the water pressure of the mainline opens the next valve. Thus, the irrigation is completed in all the group of drip laterals. Such types of systems are suitable for greenhouses, gardens, and nurseries.

11.1.2 Volume-based Sequential Hydraulically Automated Irrigation System

This type of system is suitable for irrigating orchards and field crops and any pipe diameter ranging from 50 to 254 mm and handles relatively large flow of water. The system can operate on a mainline and automatically opens the valves on the laterals. The system consists of hydraulic automatic metering valves, with or without an arrangement for cumulative readings of the quantity of water discharged (Fig. 11.2). Each metering valve can operate as an independent unit or can also be used to activate a number of supplementary hydraulic valves. The discharge of such units depends on the maximum quantity of water, which may pass through the line without any excessive friction losses.

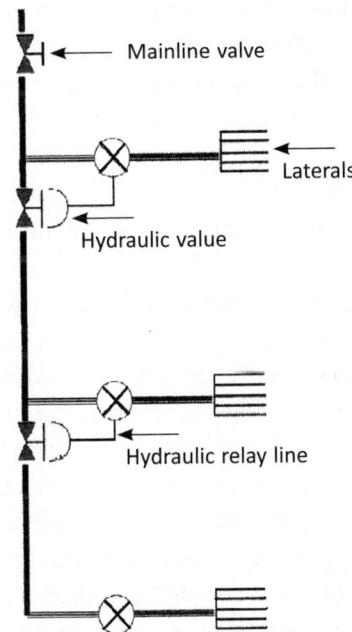

Fig. 11.2: Volume based sequential hydraulic system

At the start of irrigation cycle, all the metering valves are pre-set to the required quantities of water. The water flow reaches the first metering valve after opening of the main valve. The water pressure is transmitted from the first metering valve through the hydraulic tubing and keeps the second metering valve closed. When the desired quantity of water has been delivered to the first sub-plot, the first metering valve closes. The pressure on the second metering valve is released and the water pressure on the main supply line opens the next valve. This cycle is repeated until the last lateral is opened and the entire irrigation cycle is completed.

11.1.3 Time-based Electrically Operated Sequential System

Sequential electrically operated systems operate in sequence to control irrigation. The remotely placed solenoid valves with electric current through cables control water application. In this system, the water quantity delivered to the different plots is regulated by timer clock. The timer clock is programmed to start and stop at desired time by the user. These types of controllers are usually designed with calendar programs so that the watering cycle can be automatically started on the desired day of the week.

This system was developed mainly for the irrigation of domestic gardens and sprinkler irrigation but in principle it can be used for any kind of permanent irrigation. Regular solenoid valves are used for handling the low-pressure discharge. Sequential hydraulically-electrically operated system uses solenoid valves to activate hydraulic valves and the overall operation is hydraulic-electric. Rest of its operation is similar to sequential electrically operated system. Such types of systems are used for large diameter pipes.

11.1.4 Non-sequential Systems

Electrically-controlled non-sequential systems are automatic to a greater extent as compared to that of sequential systems. These systems control electric or hydraulic valves, which operate independently in terms of the quantity of water to be applied or frequency of irrigation. Water in each unit is supplied in different quantity and valve opens at a different time in response to a pre-determined program or to soil water content. The control panel contains electrical circuits to operate the pump or main valve, to add fertilizer according to a pre-set schedule, and to measure soil moisture, so that irrigation can be supplied to meet crop need. Such systems are usually remote-controlled, and are designed for feedback of the data received from the field, so that automatic regulation can be taken up and adjustments for changes in the pressure and discharge rate of the supply line can be accommodated. Non-sequential systems are further classified into two categories, i.e. (i) Feedback control, (ii) Inferential control.

11.1.4.1 Feedback Control System

The main components of the feedback control system are soil moisture sensors, electronic control unit such as comparators, microprocessor, computer, and solenoid valves. A farm is divided into different sub-plots. Each sub-plot has one soil moisture sensor. The sensors are connected through proper interfacing circuit to the central control unit. The central control unit is a microprocessor or a computer which can be instructed to take up a program to make irrigation scheduling decisions or comparator circuit (Fig. 11.3). Based on the soil moisture status of a sub-plot, controller decides when to start and stop the irrigation. The electronic controller actuates and de-actuates solenoid valves to start and stop the irrigation through proper interfacing circuit and relay switches.

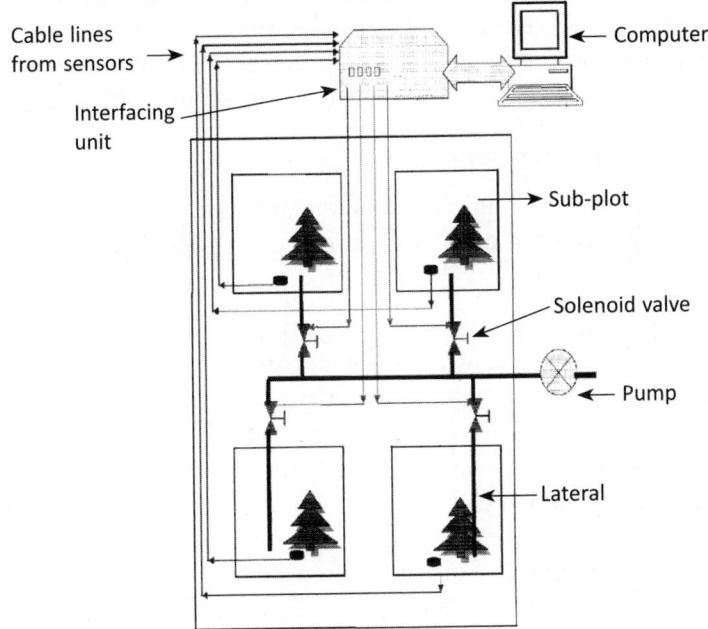

Fig. 11.3: Components of computer controlled irrigation system and their layout

11.1.4.2 Inferential Control System

This system is basically based on available evapotranspiration models to determine evapotranspiration of crop using weather parameters. The estimated evapotranspiration is used to schedule irrigation. The main parts of inferential control system are sensors for measuring various weather, evapotranspiration model, and solenoid valves. The solenoid valves are operated using interfacing circuit and relay switches to put on or to put off the irrigation system.

11.2 Components of Automated Irrigation System

An automated irrigation comprises components such as sensors, communication lines, controllers and actuators. Sometime, fertigation units, filters and back flush device for filters have also automated operation with drip irrigation system. The description of different types of soil moisture sensors, communication cables, controllers and actuators are briefly presented in the following section.

11.2.1 Irrigation Controllers

This is one of the most important components of an automated irrigation system. It turns the irrigation system on and off at the pre-determined times. In other words, the controller controls the irrigation system and the operator controls the controller. The right controller may cause the considerable water savings and lower energy bills. There may be one separate valve that controls the flow of water to a specific crop. The controller can be programmed to determine when, how frequent, and how long each valve is open. The controller should have programming flexibility so that water can be applied more efficiently.

11.2.2 Soil Moisture Sensors

Soil moisture sensors measure the water content in the soil. Soil moisture sensors commonly used for automation purpose are tensiometer and resistance block type.

Tensiometer

The tensiometer is a device, which provides direct measure of tenacity with which water is held by the soil. Tensiometers are commercially available in numerous configurations. The main disadvantage of the tensiometer is that it functions only from 0.0 to about −0.8 bar, which represents a small part of the entire range of available water. The lower moisture limit for the good crop growth is beyond the tensiometer range. Any change in soil water causes corresponding change in soil moisture tension. In automated irrigation systems, the tensiometer is modified to read change in soil moisture tension in terms of change in voltage or resistance.

Resistance Blocks

In this type of sensor, the electrical resistance between the two electrodes varies with moisture of resistance block, which is in equilibrium with the soil moisture in the crop root zone. The presence of salt or salinity in irrigation water or soil affects the observations. The gypsum block and granular matrix type sensors are commonly used for soil moisture sensing.

Gypsum Block

One of the most common methods of estimating matric potential is with gypsum or porous blocks. The device consists of a porous block containing two electrodes connected to a wire lead. The porous block is made of gypsum. When the device is buried in the soil, water will move in or out of the block until the matric potential of the block and the soil are the same. The electrical conductivity of the block is then observed. A calibration curve is made to relate electrical conductivity to the matric potential for any particular soil. The disadvantage of the porous block system is that each block has different characteristics and must be individually calibrated. Gypsum blocks are easy to use and economical but the inherent disadvantage with this sensor is that gypsum dissolves with water, calibration curve changes with the time in the same location.

Granular Matrix

These sensors reduce the problems inherent in gypsum blocks by use of a granular matrix which is supported in a metal or plastic screen. The electrodes inside the sensors are embedded in the granular fill material above the gypsum wafer. The gypsum wafer slowly neutralizes the salinity of the soil solution, hence electrical resistance between the electrodes is unaffected. Particle size of the granular fill material and its compression determine the pore size distribution in sensor. Such sensors require little maintenance during the growing season and suited for sensing soil water potential and automatic control of irrigation systems. They have advantages of low unit cost and simple installation procedures similar to those used for tensiometers.

Sensors for Climatological Parameters

Sensors for measuring various climatological parameters such as solar radiation, maximum and minimum temperature, wind speed, relative humidity, pan evaporation, etc. are interfaced to the computer to estimate evapotranspiration of crop and irrigation is commissioned based on evapotranspiration demand. Sensor to measure evaporation from a pan is interfaced with irrigation controller to actuate and deactuate the irrigation system.

An automation of irrigation systems has several positive effects. Once installed, the water distribution on fields or small-scale gardens is easier and does not have to be permanently controlled by an operator. There are several solutions to design automated irrigation systems. Modern big-scale systems allow big areas to be managed by one operator only. Sprinkler, drip or sub-surface drip irrigation systems require pumps and some hightech-components and if used for large surfaces, skilled operators are also required. Extremely high-tech solutions also exist using GIS and satellites to automatically measure the water needs content of each crop parcel and optimise the irrigation system. Drip Irrigation System can also be automated based on GSM system and android. The connections between the two mobiles are done using GSM. The GSM modules and microcontroller are connected using universal asynchronous receiver/transmitter (UART).

When the soil moisture sensor identifies the moisture content of the soil beyond a certain level, it sends a signal to the microcontroller which in turn sends a signal to a designated mobile. The mobile automatically activates the buzzer indicating the valve to be opened. When button is pressed in the called function, the signal is sent to the microcontroller which further sends the signals to the valves and finally the valve gets opened which starts the pump. The water is provided to the root of the plant drop-by-drop, and when the soil moisture becomes sufficient (field capacity), the soil moisture sensor analyzes the moisture level and sends back the signal to the microcontroller and the buzzer becomes off. After that by pressing the button in the calling function, the valve is closed (Tupe et al., 2015).

Many initiatives have been taken in this direction, but it will take time to reach to the farmers, but the system can be installed in the field of large growers who are producing cash crops in large areas or practicing the agriculture professionally. Irrigation automation is justified where a large irrigated area is divided into small segments called irrigation blocks and segments are irrigated in sequence to match the discharge available from the water source.

■■

BIBLIOGRAPHY

1. Ah Koon, P. D., Gregory, P. J. and Bell, J.P. Influence of drip irrigation emission rate on distribution and drainage of water beneath a sugar cane and a fallow plot. Agric. Water Manage.1990; 17:267–282.

2. Allen, R. G., Pereira, L., Raes, D. and Smith, M. 1998. Crop Evapotranspiration: Guidelines for Computing Crop Water Requirements. FAO Irrigation and Drainage Paper 56. Rome, Italy: Food and Agriculture Organisation.

3. American Society of Agricultural Engineers. 1985. ASAE Engineering Practice, ASAE EP405.

4. American Society of Civil Engineers. 1990. Evapotranspiration and Irrigation Water Requirements: a Manual, Committee on Irrigation Water Requirements of the Irrigation and Drainage Division of the ASCE. 332.

5. Amin, M.S.M. 1990. Hydraulic analysis of trickle lateral. Ph.D. thesis, Southampton University, UK.

6. Angelakis, A.N., Kadir, T.N. and Rolston, D.E. Time-dependent soil-water distribution under a circular trickler source. Water Management, 1993; 7:225-235.

7. Anonymous. Code for Design and Installation of Trickle Irrigation System, Indian Standard Code IS: 10799. 1984; (Part 1): 1–15.

8. Anwar, A.A. Factor G for pipelines with equally spaced multiple outlets and outflow. J. Irrig. and Drain. Engrg., ASCE, 1999; 125(1):34–38.

9. Assouline, S. Infiltration into soils: Conceptual approaches and solutions. Water Resour. Res. 2013; 49: 1755–1772.

10. Assouline, S., Cohen, S., Meerbach, D., Harodi, T. and Rosner, M. Micro-drip irrigation of field crops: Effect on yield, water uptake, and drainage in sweet corn. Soil Sci. Soc. Am. J. 2002; 66:228–235.

11. Ben-Asher, J., Charach, Ch. and Zemel, A. 1986. Infiltration and Water Extraction from Trickle Irrigation Source: The Effective Hemisphere Model1. Soil Science Society of America Journal, 50. DOI:10.2136/sssaj1986.03615995005000040010x.

12. Bernstein, L. and Francois, L.E. Leaching requirement studies: Sensitivity of alfalfa to salinity of irrigation and drainage water. Soil Sci. Soc. Amer. Proc., 1973; 37:931–943.

13. Blaney, H.F. and Criddle, W.D. 1950. Determining Water Requirements in Irrigated Areas from Climatological Irrigation Data. Technical Paper No. 96, US Department of Agriculture, Soil Conservation Service, Washington, D.C., 48.

14. Blass, S. 1964. Sub-surface irrigation. Hassadeh, 45: 1 (in Hebrew).

15. Bowen, I.S. The ratio of heat losses by conduction and by evaporation from any water surface. Phys. Rev. 1926; 27:779–787.

16. Bralts, V.F. and Edwards, D.M. Field evaluation of sub-main units. Trans. Am. Soc. Agric. Eng. 1986; 29 (6): 1659–1664.

17. Bralts, V.F. and Segerlind, L.J. Finite elements analysis of drip irrigation submains units. ASAE, 1985; 28(3):809–814.

18. Bralts, V.F., Kelly, S.F., Shayya, W.H. and Segerlind, L.J. Finite elements analysis of micro-irrigation hydraulics using virtual emitter systems. ASAE, 1993; 36 (3):717–725.

19. Bralts, V.F., Shayya, W.H., Driscoll, M.A., Cao, L. 1991. An expert system for the hydraulic design of microirrigation systems. International Summer Meeting, ASAE, New Mexico, Paper No. 91-2153, 12 p.

20. Burt, C., O'Connor, K. and Ruehr, T. 1995. Fertigation. San Luis Obispo: California Polytechnic State University.

21. Cardon, E.G. and Letey, J. Soil-based irrigation and salinity management models: II, water and solute movement calculations. Soil Sci. Soc. Am. J. 1992b; 56:1887–1892.

22. Celia, M.A., Bouloutas, E.T. and Zarba, R.L. A general mass-conservative numerical solution for the unsaturated flow equation. Water Resour. Res. 1990; 26(7): 1483–1496.

23. Christiansen, J.E. Hydraulics of sprinkling systems for irrigation. ASCE, 1942; 107: 221–239.

24. Clapp, R.B. and Hornberger, G.M. Empirical equations for soil hydraulic properties. Water Resources Research, 1978; 14: 601–604.

25. Coelho, E.F. and Or, D. A parametric model for two-dimensional water uptake by corn roots under drip irrigation. Soil Sci. Soc. Am. J., 1996; 60:1039–1049.

26. Coelho, E.F. and Or, D. Root distribution and water uptake patterns of corn under surface and subsurface drip irrigation. Plant Soil, 1999; 206:123–136.

27. Cook, F.J., Thorburn, P.J., Fitch, P. and Bristow, K.L. WetUp: A software tool to display approximate wetting patterns from drippers. Irrig. Sci., 2003; 22:129–134.

28. Cote, C.M., Bristow, K.L., Charlesworth, P.B., Cook, F.J. and Thorburn, P.J. Analysis of soil wetting and solute transport in subsurface trickle irrigation. Irr. Sci., 2003; 22:143–156.

29. Dalton, J. Experimental essays on the constitution of mixed gases: on the force of steam or vapour from water or other liquids in different temperatures, both in a Torricelli vacuum and in air; on evaporation; and on expansion of gases by heat. Manchester Lit. Phil. Soc. Mem. Proc., 1802; 5:536–602.

30. Davis, S. 1974. History of drip irrigation. Agribusiness News 10(7):1.

31. Farthing, M.W. and Ogden, F.L. 2017. Numerical Solution of Richards' Equation: A Review of Advances and Challenges, Soil Science Society of America Journal, 81:1257–1269. DOI: 10.2136/sssaj2017.02.0058.

32. Hammami, M., Hedi, D., Jelloul, B. and Mohamed, M. Approach for predicting the wetting front depth beneath a surface point source: Theory and numerical aspect. Irrig and Drain, 2002; 51:347-360.

33. Hanson, B., and May, D. 2011. Drip Irrigation Salinity Management for Row Crops, Agricultural and Natural Resources, University of California, Publication No. 8447.

34. Hargreaves, G.H. and Samani, Z.A. Estimating PET. Tech Note, J. Irrig. Drain Eng., 1982; 108(3):225–230.

35. Hargreaves, G.H. and Samani, Z.A. Reference Crop Evapotranspiration from Temperature. Appl. Eng. Agric., 1985; 1(2):96–99.

36. Hartz, T.K., Smith, R.F. LeStrange, M. and Schulbach, K.F. On-farm monitoring of soil and crop nitrogen status by nitrate-selective electrode. Commun. Soil Sci. Plant Anal., 1993; 24:2607–2615.

37. Hathoot, H.M., Al-Amoud, A.I. and Mohammad, F.S. Analysis and design of trickle irrigation laterals. American Society of Civil Engineers. Journal of Irrigation and Drainage Division, 1993; 119(5):756–767.

38. Helmi, M.H., Ahmed, I.A.A. and Fawzi, S.M. Analysis and design of trickle irrigation laterals. ASCE, 1993; 119 (5):756–767.

39. Hillel, D. Introduction to soil physics, Academic, San Diego, 1982. Chapters 1–7, 1–134.

40. Hills, D.J. and Povoa, A.F. 1993. Pressure sensivity to microirrigation emitter plugging. International Summer Meeting, ASAE/CSAE, Washington. Paper No. 93-2130, 17 p.

41. Hochmuth, G.J. Sufficiency ranges for nitrate-nitrogen and potassium for vegetable petiole sap quick tests. Hort Technology, 1994; 4:218–222.

42. Howell, T.A. and Barinas, F.A. Pressure losses across trickle irrigation fittings and emitters. Trans ASAE, 1980; 23(4):928–933.

43. Howell, T.A. and Hiler, E.A. Trickle irrigation lateral design. ASAE, 1974; 17 (5):902–908.

44. Howell, T.A. and Holer, E.A. Designing trickle irrigation laterals for uniformity. Journal of Irrigation and Drainage, Div. ASCE, 1974; 100(IR4): 443–445.

45. Howell, T.A., Stevenson, D.S., Aljibury, F.K., Gitlin, H.M., Wu, J.P., Warrick, A.W. and Raats, P.A.C. 1982. Design and operation of trickle (drip) system. In Design and Operation of Farm Irrigation System, ed. M.E. Jensen, Chpt. 16. ASAE Monograph No.3.

46. James, G.L. 1988. Principles of Farm Irrigation System Design. John Wiley & Sons, Inc. USA.

47. Jensen, M.C. and Fratini, A.M. Adjusted 'F' factors for sprinkler lateral design. Agric. Engrg., 1957; 38(4):247.

48. Jensen, M.E. 1980. Design and Operation of Farm Irrigation Systems. ASAE Monograph 3, American Society of Agricultural Engineers. St. Joseph, Michigan.

49. Kang, Y. and Nishiyama, S. Finite element method analysis of micro-irrigation system pressure distribution. Trans., Japanese Soc. of Irrig., Drain., and Reclamation Eng., Tokyo, Japan, 1994; 169:19–26.

50. Kang, Y. and Nishiyama, S. Analysis and design of micro-irrigation laterals. American Society of Civil Engineers. Journal of Irrigation and Drainage Division. 1996a; 122(2):75–81.

51. Kang, Y. and Nishiyama, S. Design of micro-irrigation sub-main units. American Society of Civil Engineers. Journal of Irrigation and Drainage Division 1996b; 122(2): 83–89.

52. Karmeli, D. and Keller, J. 1975. Trickle Irrigation Design. Glendora, California: Rain Bird Sprinkler Manufacturing Corp.

53. Keller, J. and Bliesner, Ron D. (Editors). 1990. Sprinkler and Trickle Irrigation. Van Nostrand Reinhold, New York.

54. Keller, J. and Karmeli, D. Trickle irrigation design parameters. ASAE, 1974; 17 (4):678–684.

55. Koenig, E. Methods of micro-irrigation with very small discharges and particularly low application rates. (In Hebrew.) Water Irrig. 1997; 365:32–38.

56. Lakhdar, Z. and Dalila, S. Analysis and design of a microirrigation lateral. Journal Central European Agriculture, 2006; 7 (1):57–62.

57. Li, J., Zhang, J. and Ren, L. Water and nitrogen distribution as affected by fertigation of ammonium nitrate from a point source. Irrig. Sci., 2003; 22:1:12–30.

58. Martin, E.C. 2009. Methods of Measuring for Irrigation Scheduling - WHEN. AZ1220, Arizona Water Series No.30.

59. Meyer, A.F. Computing runoff from rainfall and other physical data, Trans. Am. Soc. Civ. Eng., 1915; 79:1055–1155.

60. Mizyed, N. 1997. Comparative analysis of techniques for solving the hydraulics of pressurized irrigation pipe networks. An-Najah University. Journal for Research (Natural Sciences) Nablus, Palestine Authority.

61. Mohammed, A.I. 2010. TIWFC- Trickle Irrigation Wetting Front Pattern Calculator. Fourteenth International Water Technology Conference, IWTC 14 2010, Cairo, Egypt.

62. Monteith, J.L. Gas exchange in plant communities. In: Evans, L. T. (ed.), Environmental control of plant growth, (1963); 95–112. Academic Press, New York.

63. Montleith, J.L. Evaporation and environment. Smyp. Soc. Expt. Biol. (1965); 19: 205–234.

64. Morton, F.I. Operational estimates of areal evapotranspiration and their significance to the science and practice of hydrology. Journal of Hydrology, 1983; 66: 1–76.

65. Mwale, S.S., Ali, A., S.N. and Sparkes, D.L. Can the PR1 capacitance probe replace the neutron probe for routine soil-water measurement? Soil Use Mgt. 2005; 21:340–347.

66. Narayanamoorthy, A. and Devika, N. Economic and Resource Impacts of Drip Method of Irrigation on Okra Cultivation: An Analysis of Field Survey Data, Journal of Land and Rural Studies, 2017; 6(1):15–33.

67. Niwas, R., Singh, D. and Rao, V.U.M. 2002. Practical manual on Evapotranspiration Estimation. CCS Haryana Agricultural University, Hissar.

68. Papadopoulos, I. Agricultural and environmental aspects of fertigation-chemigation in protected agriculture under Mediterranean and arid climates. 1993; 103–133. In

Proc. on "Environmentally Sound Water Management of Protected Agriculture under Mediterranean and Arid Climates" Bari, Italy.

69. Penman, H.L. Natural evaporation from open water, bare soil and grass, Proc. R. Soc. Lond., 1948; 193:120–145.

70. Perolt, P.R. Design of irrigation pipe laterals with multiples outlets. ASCE, 1977; 103:179–193.

71. Priestley, C.H.B. and Taylor, R.J. On the assessment of surface heat flux and evaporation using large-scale parameters. Monthly Weather Review, 1972; 100(2): 81–92.

72. Prueger, J.H., Hatfield, J.L., Aase, J.K. and Pikul, Jr. J.L. Bowen-Ratio Comparisons with Lysimeter Evapotranspiration. Agronomy Journal, 1997; 89(5):730–736.

73. Ragab, R. A holistic generic integrated approach for irrigation, crop and field management: the SALTMED model. Environ. Modell. Software 2002; 17:345–361.

74. Raine, S.R., Meyer, W.S., Rassam, D.W., Hutson, J.L. and Cook, F.J. 2005. Soil-water and salt movement associated with precision irrigation systems - Research Investment Opportunities. Final report to the National Program for Sustainable Irrigation. CRCIF Report number 3.13/1. Cooperative Research Centre for Irrigation Futures, Toowoomba.

75. Rauschkolb, R.S., Rolston, D.E., Miller, R.J., Carlton, A.B. and Burau, R.G. Phosphorus fertilization with drip irrigation. Soil Sci. Soc. Am. J. 1976; 40:68–72.

76. Reeve, R.C. and Fireman, M. Salt problems in relation to irrigation. In Hagan, R.M., Haise H.R. and Edminster T.W. Madison Wisc. USA: Am. Soc. Ag. Pub., Irrigation of Agricultural lands, 1967; 998–999.

77. Scaloppi, E.J. 1988. Adjusted F factor for multiple outlet pipes. American Society of Civil Engineers. Journal of Irrigation and Drainage Division, 114(1):169–174.

78. Schwartzman, M. and Zur, B. 1986; Emitter spacing and geometry of the wetted soil volume. J. Irrig. Drain. Eng., 112:242–253.

79. Shedeed, S.I., Zaghloul, S.M. and Yassen, A.A. Effect of method and rate of fertilizer application under drip irrigation on yield and nutrient uptake by tomato. Ozean Journal of Applied Sciences, 2009; 2(2):139–147.

80. Shuttelworth, Evaporation, In: Maidment, D.R. (Ed.) Handbook of Hydrology. McGraw-Hill Book Company, New York, 1993; 4.1–4.53.

81. Shuttleworth, W.J. and Calder, I.R. Has the Priestley-Taylor equation any relevance to the forest evaporation? Journal of Applied Meteorology, 1979; 18: 639–646.

82. Simunek, J., van Genuchten, M. Th. and Sejna, M. 2005. The HYDRUS-1D software package for simulating the movement of water, heat, and multiple solutes in variably saturated media, version 3.0, HYDRUS software series 1, Department of Environmental Sciences, University of California Riverside, Riverside, CA.

83. Singh, R.N. and D., Rao, V.U.M. 2002. Practical manual on evapotranspiration estimation, CCS Haryana Agricultural University, Hisar, India.

84. Solomon, K. and Keller, J. Trickle irrigation uniformity and efficiency. ASCE, 1978; 104 (3):293–306.

85. Stagnitti, F., Parlange, J.Y. and Rose, C.W. Hydrology of a small wet catchment. Hydrological Processes, 1989; 3:137–150.

86. Stangheuini, C. Mixed convection above greenhouse crop canopies. Agric. For. Meteorol., 1993; 66:111-117.

87. Taghavi, S.A., Marino, M.A. and Rolston, D.E. 1984. Infiltration from trickle irrigation source. J. of Irrig. And Drainage Eng., ASCE 11094.

88. Thorburn, P.J., Cook, F.J. and Bristow, K.L. 2002. New water-saving production technologies: Advances in trickle irrigation. In Water for Sustainable Agriculture in Developing Regions – More Crop from Every Scarce Drop (Eds M. Yajima, K. Okada and N. Matsumoto) Proceedings 8th JIRCAS International Symposium, Tsukuba, Japan, 27–28 November: 53–62.

89. Thornthwaite, C.W. An approach toward a rational classification of Climate. Geograph. Rev. 1948; 38 (1):55–94.

90. Thornthwaite, C.W. and Holzman, B. The determination of land and water surfaces, Month. Weather Rev., 1939; 67:4–11.

91. Thornthwaite, C.W. and Mather, J.R. 1955. The Water Balance. Publications in Climatology, Drexel Institute of Technology, Centerton, New Jersey, Vol. VIII, No. 1.

92. Tupe, A.R., Gaikwad, A.A. and Kamble, S.U. Intelligent Drip Irrigation System International Journal of Innovative Research in Advanced Engineering (IJIRAE), 2015; 2(2):120–125.

93. USDA-NRC. (1984). National Engineering Handbook. Section 15. Furrow Irrigation. National Technical Information Service, Washington, DC, Chapter 5.

94. Valiantzas, J.D. Analytical approach for direct drip lateral hydraulic calculation. ASCE, 1998; 124 (6):300–305.

95. van Bavel, C.H.M. Potential evaporation; the combination concept and its experimental verification. Water Resources Res. 1966; 2 (3): 455–467.

96. Verburg, K., Bridge, B.J., Bristow, K.L. and Keating, B.A. 2001. Properties of selected soils in the Gooburrum – Moore Park area of Bundaberg. - Technical Report 09/01, CSIRO Land and Water, Canberra, Australia.

97. Vermeiren, L. and Jobling, G. A. 1984. Localized irrigation. FAO Irrigation and Drainage Paper No. 36.

98. Vitosh, M.L. and Silva, G.H. A rapid petiole sap nitrate-nitrogen test for potatoes. Commun. Soil Sci. Plant Anal. 1994; 25:183–190.

99. Watters, G.Z. and Keller, J. 1978. Trickle irrigation tubing hydraulics. ASAE Paper No., 78-2015. St. Joseph MI: ASAE.

100. Wu, I.P. and Gitlin, H.M. Drip irrigation based on uniformity. ASAE. 1974; 3:429–432.

101. Yitayew, M. and Warrick, A.W. Trickle lateral hydraulics II: Design and examples. ASCE, 1988; 114(2):289–300.

102. Zella, L. and Kettab, A. Numerical methods of micro-irrigation lateral design. Biotechnol. Agron. Soc Environ. 2002; 6 (4):231–235.

103. Zella, L., Kettab, A. and Chasseriaux, G. Design of a micro-irrigation system based on the control volume method. Biotechnol. Agron. Soc. Environ. 2006; 10(3):163–171.

APPENDICES

Appendix A

Pan coefficient for Class A pan for different grounds and level of mean relative humidity (RH) and 24-hour wind run (adopted from Ram Niwas et al., 2002)

Wind	Windward distance (m)	Pan placed in short green cropped area			Pan placed in dry fallow area		
		RH mean			RH mean		
		Low	Medium	High	Low	Medium	High
Light	1	0.55	0.65	0.75	0.70	0.80	0.85
	10	0.65	0.75	0.85	0.60	0.70	0.80
	100	0.70	0.80	0.85	0.55	0.65	0.75
	1000	0.75	0.85	0.85	0.50	0.60	0.70
Moderate	1	0.50	0.60	0.65	0.65	0.75	0.80
	10	0.60	0.70	0.75	0.55	0.65	0.70
	100	0.65	0.75	0.80	0.50	0.60	0.65
	1000	0.70	0.80	0.80	0.45	0.55	0.60
Strong	1	0.45	0.50	0.60	0.60	0.65	0.70
	10	0.55	0.60	0.65	0.50	0.55	0.65
	100	0.60	0.65	0.70	0.45	0.50	0.60
	1000	0.65	0.70	0.75	0.40	0.45	0.55
V. strong	1	0.40	0.45	0.50	0.50	0.60	0.65
	10	0.45	0.55	0.60	0.45	0.50	0.55
	100	0.50	0.60	0.65	0.40	0.45	0.50
	1000	0.55	0.60	0.65	0.35	0.40	0.45

RH mean-low <40%, medium 40–70%, high >70%:

Wind-light <175 m/day, moderate 175–425 m/day, strong 425–700 m/day:

very strong >700 m/day

Appendix B

Crop coefficients (K_c) of some field crops

Crops	Stages					Whole crop season
	Initial	Vegetative	Mid season	Late season	Harvest stage	
Cotton	0.45	0.75	1.15	0.85	0.68	0.85
Groundnut	0.45	0.75	1.03	0.80	0.58	0.78
Maize	0.40	0.78	1.13	0.88	0.75	0.88
Rice	1.13	1.30	1.12	1.00	1.00	1.13
Sorghum	0.35	0.73	1.13	0.78	0.58	0.80
Soybean	0.35	0.75	1.13	0.75	0.45	0.83
Sugarcane	0.45	0.85	1.15	0.78	0.55	0.95
Sunflower	0.35	0.75	1.13	0.75	0.40	0.80
Wheat	0.35	0.75	1.13	0.70	0.23	0.85
Tomato	0.45	0.75	1.13	0.88	0.63	0.83
Cabbage	0.45	0.75	1.03	0.95	0.88	0.75

Appendix C

Monthly values of '*i*' corresponding to mean monthly temp in °C for use in the Thornwaite formula

Temp	0.0	0.3	0.6	0.9	Temp	0.0	0.6	0.9
5	1.00	1.09	1.19	1.29	23	10.08	10.48	10.68
6	1.32	1.42	1.52	1.63	24	10.75	11.16	11.37
7	1.66	1.77	1.89	2.00	25	11.44	11.85	12.06
8	2.04	2.15	2.27	2.39	26	12.13	12.56	12.78
9	2.44	2.56	2.69	2.81	27	12.85	13.28	13.50
10	2.86	2.99	3.12	3.25	28	13.58	14.02	14.24
11	3.30	3.44	3.58	3.72	29	14.32	14.77	14.99
12	3.76	3.91	4.05	4.20	30	15.07	15.33	15.76
13	4.25	4.40	4.55	4.70	31	15.84	16.30	16.54
14	4.75	4.91	5.07	5.22	32	16.62	17.09	17.33
15	5.28	5.44	5.60	5.76	33	17.41	17.89	18.13
16	5.80	5.98	6.15	6.32				
17	6.37	6.55	6.72	6.89				
18	6.95	7.13	7.30	7.48				
19	7.54	7.72	7.91	8.09				
20	8.15	8.34	8.53	8.72				
21	8.78	8.97	9.16	9.30				
22	9.42	9.47	9.81	10				

Appendix D

Ratio of slope of the saturation vapour pressure to the psychometric constant (s/y) in terms of air temperature (°C) in tenths

Temp	0.0	0.10	0.20	0.30	0.40	0.50	0.60	0.70	0.80	0.90
10	1.258	1.265	1.272	1.279	1.286	1.293	1.300	1.307	1.314	1.322
11	1.329	1.336	1.343	1.350	1.358	1.365	1.373	1.381	1.389	1.397
12	1.405	1.413	1.421	1.429	1.437	1.445	1.453	1.461	1.469	1.477
13	1.486	1.494	1.502	1.510	1.519	1.528	1.536	1.545	1.554	1.563
14	1.572	1.581	1.590	1.599	1.608	1.617	1.626	1.635	1.644	1.653
15	1.663	1.672	1.681	1.690	1.700	1.710	1.719	1.729	1.739	1.749
16	1.759	1.769	1.779	1.789	1.799	1.810	1.820	1.830	1.840	1.850
17	1.860	1.871	1.881	1.891	1.902	1.913	1.924	1.934	1.945	1.955
18	1.966	1.977	1.988	1.999	2.010	2.022	2.033	2.044	2.055	2.066
19	2.077	2.089	2.100	2.112	2.124	2.136	2.174	2.159	2.171	2.182
20	2.194	2.206	2.218	2.230	2.242	2.255	2.267	2.279	2.291	2.303
21	2.315	2.328	2.340	2.353	2.36	2.379	2.391	2.404	2.417	2.429
22	2.442	2.455	2.469	2.482	2.496	2.509	2.522	2.536	2.549	2.563
23	2.576	2.590	2.607	2.618	2.632	2.646	2.659	2.673	2.687	2.701
24	2.715	2.730	2.744	2.759	2.733	2.788	2.802	2.817	2.831	2.846
25	2.860	2.875	2.890	2.906	2.921	2.936	2.951	2.966	2.982	2.997
26	3.012	3.028	3.044	3.060	3.076	3.092	3.107	3.123	3.139	3.155
27	3.171	3.188	3.204	3.221	3.237	3.254	3.270	3.287	3.303	3.320
28	3.336	3.353	3.371	3.338	3.405	3.423	3.440	3.457	3.473	3.452
29	3.509	3.527	3.545	3.563	3.581	3.599	3.617	3.635	3.653	3.671
30	3.689	3.708	3.727	3.746	3.765	3.784	3.802	3.821	3.840	3.859
31	3.878	3.898	3.917	3.937	3.956	3.976	3.996	4.015	4.035	4.054
32	4.074	4.094	4.115	4.135	4.156	4.176	4.196	4.217	4.237	4.258
33	4.278	4.299	4.320	4.342	4.363	4.384	4.405	4.426	4.448	4.469
34	4.490	4.512	4.534	4.556	4.578	4.660	4.622	4.644	4.666	4.668
35	4.712	4.735	4.758	7.781	4.804	4.827	4.850	4.873	4.896	4.919
36	4.943	4.967	4.991	5.015	5.039	5.063	5.087	5.111	5.135	5.159
37	5.184	5.209	5.234	5.259	5.284	5.309	5.334	5.352	5.384	5.409
38	5.435	5.461	5.487	5.513	5.539	5.564	5.591	5.614	5.640	5.678

Appendix E

Crop coefficient for use in Blaney-Criddle formula

Months	Crops							
	Rice	*Maize*	*Wheat*	*S. cane*	*Cotton*	*Vegetable*	*Berseam*	*Citrus*
Jan	-	-	0.50	0.75	-	0.50	0.50	0.50
Feb	-	-	0.70	0.80	-	0.55	0.70	0.55
Mar	-	-	0.75	0.85	-	0.60	0.80	0.55
Apr	0.85	0.50	0.70	0.85	0.50	0.65	0.90	0.60
May	1.00	0.60	-	0.90	0.60	0.70	1.00	0.60
Jun	1.15	0.70	-	0.95	0.75	0.75	-	0.65
Jul	1.30	0.80	-	1.00	0.90	0.80	-	0.70
Aug	1.25	0.80	-	1.00	0.85	0.80	-	0.70
Sep	1.10	0.60	-	0.95	0.75	0.70	0.60	0.65
Oct	0.90	0.50	0.70	0.90	0.55	0.60	0.65	0.60
Nov	-	-	0.65	0.85	0.50	0.55	0.70	0.55
Dec	-	-	0.60	0.75	0.50	0.50	0.60	0.55

Appendix F

Value of σT^4 expressed as mm of water per day for different temperatures, °C

Temp.	0.0	0.1	0.2	0.3	0.4	0.5	0.6	0.7	0.8	0.9
10.0	12.83	12.85	12.87	12.89	12.91	12.93	12.95	12.97	12.99	13.01
11.0	13.02	13.04	13.06	13.08	13.10	13.12	13.14	13.16	13.18	13.20
12.0	13.22	13.24	13.26	13.28	13.30	13.32	13.34	13.36	13.38	13.40
13.0	13.42	13.44	13.46	13.48	13.50	13.52	13.54	13.56	13.58	13.60
14.0	13.62	13.64	13.66	13.68	13.70	13.72	13.74	13.76	13.78	13.81
15.0	13.83	13.85	13.86	13.88	13.90	13.92	18.94	13.96	13.98	14.01
16.0	14.03	14.05	14.07	14.09	14.11	14.13	14.15	14.17	14.19	14.21
17.0	14.23	14.25	14.27	14.29	14.32	14.34	14.36	14.38	14.40	14.42
18.0	14.44	14.46	14.48	14.51	14.53	14.55	14.57	14.59	14.61	14.63
19.0	14.65	14.67	14.70	14.72	14.74	14.76	14.78	14.81	14.83	14.85
20.0	14.87	14.89	14.91	14.93	14.96	14.98	15.00	15.02	15.04	15.07
21.0	15.09	15.11	15.15	15.15	15.17	15.19	15.22	15.24	15.26	15.29
22.0	15.31	15.33	15.35	15.37	15.39	15.41	15.44	15.46	15.48	15.51
23.0	15.53	15.55	15.57	15.59	15.62	15.64	15.66	15.69	15.71	15.73
24.0	15.75	15.77	15.30	15.32	15.84	15.87	15.89	15.91	15.93	15.95
25.0	15.98	16.01	16.03	16.05	16.07	16.09	16.12	16.14	16.16	16.19
26.0	16.21	16.23	16.26	16.28	16.30	16.33	16.35	16.37	16.39	16.41
27.0	16.44	16.47	16.49	16.51	16.53	16.55	16.58	16.81	16.63	16.65
28.0	16.68	16.47	16.72	16.75	16.77	16.79	16.82	16.85	16.87	16.89
29.0	16.91	16.93	16.96	16.99	17.01	17.03	17.06	17.09	17.11	17.13
30.0	17.15	17.17	17.20	17.23	17.25	17.27	17.30	17.33	17.35	17.37
31.0	17.40	17.43	17.45	17.47	17.49	17.51	17.54	17.57	17.59	17.61
32.0	17.64	17.67	17.69	17.71	17.74	17.77	17.79	17.81	17.84	17.87
33.0	17.89	17.91	17.94	17.97	17.99	18.01	18.04	18.07	18.09	18.11
34.0	18.14	18.17	18.19	18.21	18.24	18.27	18.29	18.31	18.34	18.37
35.0	18.39	18.42	18.45	18.47	18.50	18.53	18.55	18.57	18.60	18.63
36.0	18.65	18.67	18.70	18.73	18.75	18.78	18.81	18.83	18.86	18.89
37.0	18.91	18.93	18.96	18.99	19.01	19.04	19.07	19.09	19.12	19.15
38.0	19.17	19.20	19.23	19.25	19.28	19.31	19.33	19.35	19.38	19.41
39.0	19.44	19.47	19.49	19.52	19.54	19.57	19.60	19.62	19.65	19.67
40.0	19.70	19.73	19.76	19.78	19.81	19.84	19.87	19.89	19.92	19.95

INDEX